Alkylphosphocholines: New Drugs in Cancer Therapy

Progress in Experimental Tumor Research

Vol. 34

KARGER

Basel · München · Paris · London · New York · New Delhi · Bangkok · Singapore · Tokyo · Sydney

Alkylphosphocholines: New Drugs in Cancer Therapy

Volume Editors
H. Eibl, Göttingen
P. Hilgard, Frankfurt
C. Unger, Göttingen

85 figures and 32 tables, 1992

KARGER

Basel · München · Paris · London · New York · New Delhi · Bangkok · Singapore · Tokyo · Sydney

Progress in Experimental Tumor Research

Bibliographic Indices
This publication is listed in bibliographic services, including Current Contents® and Index Medicus.

Drug Dosage
The authors and the publisher have exerted every effort to ensure that drug selection and dosage set forth in this text are in accord with current recommendations and practice at the time of publication. However, in view of ongoing research, changes in government regulations, and the constant flow of information relating to drug therapy and drug reactions, the reader is urged to check the package insert for each drug for any change in indications and dosage and for added warnings and precautions. This is particularly important when the recommended agent is a new and/or infrequently employed drug.

Printed in Switzerland on acid-free paper by Thür AG Offsetdruck, Pratteln
ISBN 3-8055-5464-8

Contents

Animal Experiments

Clinical Studies

Acknowledgments

The present work was supported by the Federal Ministry of Research and Technology (BMFT), Bonn-Bad Godesberg, and by the ASTA-Pharma AG, Frankfurt/Main, FRG.

Introduction

Alkylphosphocholines represent a new class of membrane-active agents displaying numerous biological effects which turn to become main topics in experimental cell research. The compounds have a relatively simple chemical structure. They are phosphocholine esters of aliphatic alcohols. Their physical properties are very similar to lysolecithins. However, large differences in comparison to lysolecithins and generally to phospholipids exist in the metabolic fate of these molecules within biological cells. For instance, substrate properties of alkylphosphocholines for phospholipid metabolizing enzymes are markedly reduced. Thus, in contrast to lysophospholipids substantial amounts of alkylphosphocholines can accumulate in cell membranes and thereby influence a variety of cell properties, for instance cell growth and cell differentiation.

Therefore, it is reasonable to discuss alkylphosphocholines as new and selective modulators of biological membranes separately from the effects of lysophospholipids. This is also justified by considering the chemical structure of these molecules. Alkylphosphocholines lack glycerol and by definition do not belong to the class of naturally occurring phospholipid molecules. For instance, hexadecylphosphocholine will not have biological activities ascribed to the so-called platelet-activating factor (PAF). In fact, it can even inhibit certain effects of PAF in biological systems, for instance platelet aggregation.

In 1984, when we started the first experiments with hexadecylphosphocholine, there was no interest in studies on alkylphosphocholines and especially not about the biological properties of these compounds. Surprisingly, in cell culture experiments some alkylphosphocholines showed strong antineoplastic activity. Later on, these results were confirmed and extended to animal tumor models in cooperation with Berger and Schmähl (DKFZ, Heidelberg). It was demonstrated that hexadecylphosphocholine showed strong antitumor activity against chemically induced autochthonous mammary carcinomas in vivo. This response has been proven to be superior over classical antitumor agents such as cyclophosphamide.

At this time, a cooperation between the research groups of Eibl, Nagel and Unger (Göttingen), Berger and Schmähl (Heidelberg) and Engel, Hilgard, Peukert and Stekar (Frankfurt, Asta Pharma AG) had been initiated and supported by a research program of the Bundesministerium für Forschung und Technologie. The combined effort resulted in the discovery of different biological effects induced by alkylphosphocholines. To promote an early clinical application, the research activities were concentrated on the properties of hexadecylphosphocholine. Therefore, the main part of the different contributions in this book 'Alkylphosphocholines: New Drugs in Cancer Therapy' is addressed to this subject, especially to the antineoplastic activity of hexadecylphosphocholine.

Shortly, we would like to itemize some important biological activities of hexadecylphosphocholine. It shows selective antineoplastic activity in vitro and in vivo. Moreover, induction of differentiation has been observed for leukcemic cell lines in vitro and for breast cancer in vivo. In addition, hexadecylphosphocholine can induce cytokine release and is capable of amplifying the effect of G-CSF on human bone marrow progenitor cells.

In contrast to the mode of action of drugs applied in cancer therapy which interact with the cell nucleus as target, hexadecylphosphocholine incorporates into the plasma membrane of tumor cells and is distributed within the cell via the band of intracellular membrane structures. It most probably acts via membranes. Cumulative evidence collected in the last 10 years points into the direction of an interference with mechanisms of the signal transduction of different mitogenic growth factors. In particular, a specific interference with PI-dependent signal transduction pathways is discussed and presently under experimental investigation.

At present, hexadecylphosphocholine is clinically investigated in phase II trials. The most interesting aspect is a marked antineoplastic activity of hexadecylphosphocholine in the topical treatment of skin metastases resulting from breast cancer and lymphomas.

The results presented in this publication were obtained in basic and applied clinical research in a time period of about 6 years. They show that the original observations on the biological activity of hexadecylphosphocholine resulted in a cascade of different and individual research projects. To conclude, it was useful to study alkylphosphocholines as a new group of potent and biologically active compounds in the past and it will be useful to continue these studies in the future.

Hansjörg Eibl
Peter Hilgard
Clemens Unger

Eibl H, Hilgard P, Unger C (eds): Alkylphosphocholines: New Drugs in Cancer Therapy. Prog Exp Tumor Res. Basel, Karger, 1992, vol 34, pp 1–5

Synthesis of Hexadecylphosphocholine (Miltefosine)

Hansjörg Eibl[a], *Jürgen Engel*[b]

[a]Max-Planck-Institut für biophysikalische Chemie, Göttingen;
[b]ASTA Pharma AG, Frankfurt/Main, FRG

Introduction

Hexadecylphosphocholine is a new drug in cancer therapy. Chemically, it is the phosphocholine ester of hexadecanol and has similar physical properties as lysolecithin. Mainly due to the lack of glycerol, however, the bioconversion of hexadecylphosphocholine and of other alkylphosphocholines differs strongly in comparison to lysolecithin [1].

In this chapter, we describe a simple procedure for the synthesis of hexadecylphosphocholine. The route is based on Eibl's [2] procedure, which was developed for the synthesis of phosphatidylethanolamines. Basic intermediates are hexadecylphosphoethanolamine or hexadecylphospho-N-(alkyl)-ethanolamines. These intermediates allow the preparation of hexadecylphosphocholine and of analogs with different degree of methylation or alkylation.

Materials

Hexadecanol (99%) was obtained from Fluka Chemie (Buchs, Switzerland). Phosphorus oxychloride was a product of Merck (Darmstadt, FRG) and purified by distillation. The fraction boiling between 106 and 107 °C was collected. Ethanolamine and the different N-(alkyl)-ethanolamines were purchased from Merck (Darmstadt, FRG). Solvents were from Baker (Deventer, The Netherlands) and used without further purification.

Experiments and Results

Hexadecanol is converted to hexadecylphosphocholine in a two-step procedure (Fig. 1). Step 1 includes three reactions – phosphorylation of hexadecanol with phosphorus oxychloride, ring closure of the respective

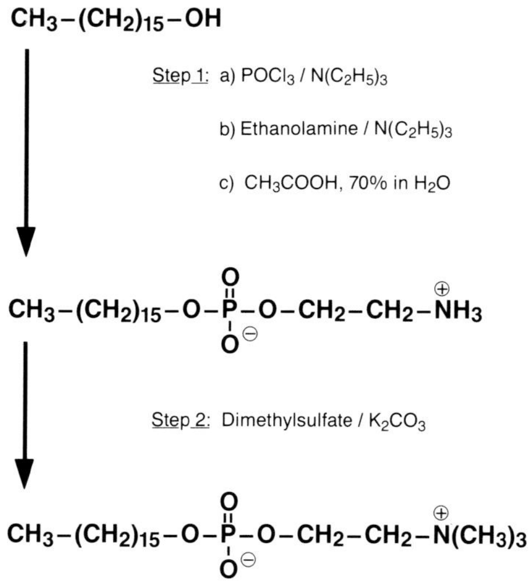

Fig. 1. Synthesis of hexadecylphosphocholine starting from hexadecanol via hexadecylphosphoethanolamine. Step 1 includes 3 chemical conversions and step 2 represents permethylation of hexadecylphosphoethanolamine.

hexadecylphosphorus dichloride with ethanolamine and selective opening of the oxazaphospholane ring at the PN bond using slightly acidic conditions, which are accomplished straightforwardly without work up of the intermediates. In step 2 permethylation of hexadecylphosphoethanolamine or of respective (N-alkyl)-ethanolamines is achieved applying a slight excess of dimethylsulfate in the presence of potassium carbonate.

Hexadecylphosphoethanolamine (Step 1)

Hexadecyldichlorophosphate. Hexadecanol (1 mol; 242 g) and triethylamine (1.1 mol; 111 g) are dissolved in 1 liter tetrahydrofuran (THF). The solution is added dropwise with vigorous stirring to a solution of phosphorous oxychloride (1.2 mol; 184 g) in 100 ml THF precooled in an ice-water bath. At no time should the temperature of the reaction mixture exceed 15 °C. After the addition of hexadecanol, the reaction mixture is kept at 20 °C for 15 min. As shown by thin-layer chromatography (TLC), the reaction is complete and no by-products can be detected. The only product formed is hexadecyldichlorophosphate (Rf value of 0.4 in diethylether after conversion to hexadecyl phosphoric acid dimethylester).

Hexadecyl-2-oxo-1,2,3-oxazaphospholane. A solution of ethanolamine (1.2 mol; 73 g) and triethylamine (2.4 mol; 243 g) in 1.5 liter THF is added dropwise under vigorous stirring to the reaction mixture of step 1a at 20 °C. The temperature of the reaction mixture is controlled by a water bath and should not exceed 40 °C. After the addition of ethanolamine and an additional 15 min at 40 °C ring closure is completed as shown by TLC (Rf value of 0.1 in diethylether). Hexadecyl-2-oxo-1,2,3-oxazaphospholane is the only product formed containing a long-chain hexadecyl residue. The reaction mixture is filtered by suction to remove precipitated triethylamine hydrochloride. The clear and slightly yellow filtrate is evaporated down. The oily residue can be used directly for further modification.

Hexadecylphosphoethanolamine. The oily residue of step 1b is dissolved in a mixture of 240 g acetic acid and of 100 ml distilled water at 70 °C. After 30 min at this temperature, ring scission at the PN bond is completed and the product is precipitated by the addition of 5 liter acetone. After cooling to 10 °C, the mixture is filtered by suction and the crystals are collected. Usually, the product is pure hexadecylphosphoethanolamine (Rf value of 0.4 in $CHCl_3/CH_3OH$/acetic acid/water 100:60:20:5 by vol). The yield is 360 g (90%) based on the starting hexadecanol. Hexadecylphosphoethanolamine; MW 365.50; elemental analysis: C, 59.15–59.01% (calculated-found); H, 11.03–10.95%; N, 3.83–3.79%; P, 8.48–8.30%.

Hexadecylphosphocholine (Step 2)

Hexadecylphosphoethanolamine (1 mol; 365 g) is dissolved in 2-propanol (1.2 liter) and dichloromethane (0.4 liter) at 40 °C (water bath). Potassium carbonate (1 mol; 138 g) in 200 ml H_2O is added and the solution is kept between 30 and 40 °C during the dropwise addition of dimethylsulfate (0.9 mol; 114 g) in 50 ml 2-propanol in a time period of 60 min. After 30 min, the N-methylation process is repeated two times with the same amounts of potassium carbonate and dimethylsulfate. Finally, an additional amount of dimethylsulfate (0.6 mol; 76 g) in 50 ml 2-propanol is added dropwise followed by the dropwise addition of potassium carbonate (0.6 mol; 83 g) in 120 ml H_2O. TLC demonstrates complete conversion of hexadecylphosphoethanolamine to hexadecylphosphocholine (Rf values of 0.4 and 0.2, respectively, in the solvent system $CHCl_3/CH_3OH$/acetic acid/H_2O 100:60:20:5 per vol). The reaction mixture is cooled to 20 °C. Two phases are formed and the lower phase is discarded. The upper phase contains the product and is evaporated to dryness. The residue is dissolved in 1 liter $CHCl_3$ at 50 °C. 2-Butanone (2 liter) is added to precipitate hexadecylphosphocholine. The crystallisation is repeated with the same amount of solvents. This usually results in pure hexadecylphosphocholine. The yield is 375 g (92%) based on hexadecylphosphoethanolamine.

Hexadecylphosphocholine · 1 H_2O; MW 407.58; elemental analysis: C, 59.27–58.98% (calculated-found); H, 11.37–11.31%; N, 3.29–3.21%; P, 7.28–7.11%.

Discussion

Many procedures for the preparation of diacylglycero-3-phosphocholines [3] and monoacyl-glycero-3-phosphocholines [4] have been described. Then, these synthetic routes have been adapted to the preparation of alkylphosphocholines, especially for the synthesis of hexadecylphosphocholine.

The widely used method of Hirt and Berchtold [5] is convenient, but suffers from low yields. In the original procedure bromoethylphosphoric acid dichloride was used as phosphorylating agent and the resulting sodium salt of hexadecylphospho bromoethylester was removed from the by-product, dihexadecylphospho bromoethylester by crystallisation. Unfortunately, amination with trimethylamine also results in considerable amounts of by-products, for instance hexadecyl-2-oxo-1,2,3-dioxaphospholane and respective hydrolysis products such as hexadecylphosphoglycerol. The yield of pure hexadecylphosphocholine prepared according to this procedure is about 45%.

An alternative route directly via the 2-oxo-1,2,3-dioxaphospholane has been first described by Phyong et al. [6] for the preparation of phosphatidylcholines and was also used for the synthesis of (alkyl)-lysolecithins and analogs [7]. 2-Chloro-2-oxo-1,3,2-dioxaphospholane is used as phosphorylating agent and the respective hexadecyl-2-oxo-1,2,3-dioxaphospholane as intermediate is converted to hexadecylphosphocholine by amination with trimethylamine. The yield is about 50% starting from hexadecanol. In both methods, the phosphorylating agents have to be prepared separately.

In our experience, the synthesis of hexadecylphosphocholine via the respective hexadecyl-2-oxo-1,2,3-oxazaphospholane described here is most convenient. The final product is obtained in two steps and phosphorus oxychloride as phosphorylating agent is successfully used without the formation of by-products. The yields are high and pure hexadecylphosphocholine is obtained without a chromatographic purification step.

The use of intermediates such as alkyl-2-oxo-1,2-3-oxazaphospholanes opens a general avenue for the preparation of different analoges of hexadecylphosphocholine (fig. 2). However, according to our experience, only small structural variations are allowed to retain the activity against autochthonous methylnitrosourea-induced mammary carcinomas. Good activity was observed for compounds with straight-chain alkyl residues R_1 between C_{16} and

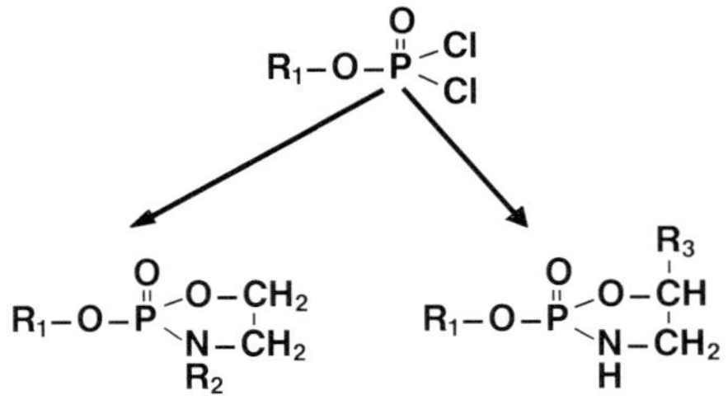

Fig. 2. General application of the synthetic procedure via oxazaphospholanes. R_1 = Dodecanol, tetradecanol, hexadecanol, octadecanol and eicosanol; R_2 = hydrogen, methyl, ethyl, propyl, butyl and pentyl; R_3 = hydrogen, methyl, ethyl and propyl. The respective alkylphosphocholines or analogues are obtained by ring cleavage and permethylation as shown in fig. 1.

C_{20} carbon atoms. An alteration of R_2 from methyl to ethyl is accepted whereas an exchange of methyl for propyl or hydrogen already results in an almost complete loss of antitumor activity. This loss of activity is also observed for an alteration of R_3 from hydrogen to methyl.

In summary, oxazaphospholanes are useful intermediates in the synthesis of phosphatidylcholines and of alkylphosphocholines. The synthetic routes are simple and allow the use of phosphorus oxychloride in the phosphorylation step. A large variety of different analogs of alkylphosphocholines can be prepared by using different ethanolamines in the ring closure reaction.

References

1 Eibl H, Unger C: Hexadecylphosphocholine, a new and selective antitumor drug. Cancer Treat Rev 1990;17:233–242.

2 Eibl H: Phospholipid synthesis: Oxazaphospholanes and dioxaphospholanes as intermediates. Proc Natl Acad Sci USA 1978;75:4074–4077.

3 Eibl H: Synthesis of glycerophospholipids. Chem Phys Lipids 1980;26:405–429.

4 Eibl H, Woolley P: A general synthetic method for enantiomerically pure ester and ether lysophospholipids. Chem Phys Lipids 1988;47:63–68.

5 Hirt R, Berchtold R: Synthesis of phospholipids. A new synthesis of lecithin. Pharm Acta Helv 1958;33:349–356.

6 Phyong NH, Thuong NT, Charbier P: New method for preparation of glycolecithins and lecithins. CR Hebd Acad Sci Paris, Ser C 1976;283:229.

7 Bosies E, Herrmann DBJ: Bicker U, Gall R, Pahlke W: Synthesis of thioether phosphocholine analoges. Lipids 1987;22:775–980.

H. Eibl, Max-Planck-Institut für biophysikalische Chemie,
Postfach 2841, D-W–3400 Göttingen (FRG)

Eibl H, Hilgard P, Unger C (eds): Alkylphosphocholines: New Drugs in Cancer Therapy. Prog Exp Tumor Res. Basel, Karger, 1992, vol 34, pp 6–11

Determination of Alkylphosphocholines and of Alkyl-glycero-phosphocholines in Biological Fluids and Tissues

Jochem Kötting[a], *Norbert W. Marschner*[b], *Clemens Unger*[b], *Hansjörg Eibl*[a]

[a]Max-Planck-Institut für biophysikalische Chemie, and [b]Medizinische Universitätsklinik, Abteilung Hämatologie/Onkologie, Göttingen, FRG

Introduction

Alkylphosphocholines (APC) are new antineoplastic agents with different biological effects. Their possible application in medicine as antitumor drugs has been proposed by Eibl and Unger [1, 2]. Due to potent antitumoral activity in vitro [3–6] and in vivo [7–8], hexadecylphosphocholine (HePC) was chosen for first clinical phase I trials.

The increasing number of laboratory and clinical studies with alkylphosphocholines and the search for a more quantitative understanding of the distribution and action of these drugs in different tissues require fast and simple analytical methods for the quantitative determination of these compounds. The first method established for analysis of HePC levels in biological fluids was a high-performance liquid chromatography assay [10]. Because of the complex and time-consuming sample work up, this method is not useful for routine determination in clinical studies. Also, a high-performance thin layer chromatography (HPTLC) method with densitometric quantitation was introduced by Rustenbeck and Lenzen [11].

In our contribution the determination of HePC is extended to structural analogs such as octadecylphosphocholine (OcPC) and erucylphosphocholine (ErPC) and also includes 1-O-octadecyl-2-O-methyl-*rac*-glycero-3-phosphocholine ($ET_{18}OCH_3$). In addition, the described assay offers an essential simplification of the extraction and isolation procedure.

Material and Methods

The synthesis of HePC is described by Eibl and Engel [this vol.]. Other procedures for the preparation of alkylphosphocholines and of $ET_{18}OCH_3$ have been described earlier

[12, 13]. Chemicals were used in the highest purity available without further purification. Solvents were purchased from Baker Chemicals (Deventer, NL), triethylamine from Fluka (Buchs, CH) and cupric sulfate and HPTLC plates silica gel 60 (200 × 100 mm) from Merck (Darmstadt, FRG).

Virgin rats of the Wistar strain were supplied from the Institut für Versuchstierkunde (Hannover, FRG). They normally had a body weight between 200 and 230 g.

For recovery experiments, 4 untreated animals were killed by CO_2 anesthesia, unbled with an infusion of 0.9% NaCl in the left ventricle and the organs were removed. For each compound 3 pieces of about 0.1–0.5 g of each organ were cut into small slices. After addition of 3 ml $CHCl_3$:MeOH:10% NaCl (2:2:1, per vol), the mixture was homogenized at 3,000 rpm in an ice-water bath by application of 10 strokes in an Elvehjem-Potter with a Teflon pestle driven by a Homogen homogenizer from Schütt (Göttingen, FRG). Sticky material from the pestle and the homogenisator was washed to the other material with 2 ml of water. Each of these 3 homogenates was spiked with an aqueous solution (1 nmol/μl) of the respective APC to a tissue concentration of 50 nmol/g. These mixtures were extracted and analyzed as described below.

For extraction, the tissue homogenates and serum samples (up to 0.5 ml) were mixed with 2 ml $CHCl_3$:MeOH (2:1, v/v) and vortexed for 30 s. For phase separation the tubes were spun at 3,000 *g* for 5 min in a centrifuge from Sigma Christ (Osterode, FRG). The proteins formed a precipitate at the interface. The lower chloroform layer contained the APGs together with membrane lipids and was transferred with a pasteur pipette to a second set of test tubes. Extraction of the upper protein/water layer was repeated two times with 2 ml $CHCl_3$:MeOH (3:1, v/v) and 2 ml $CHCl_3$. The chloroform extracts were combined and the solvent removed carefully in a sand bath (50 °C) under a gentle stream of nitrogen. Depending on the APC concentration expected in the samples they were dissolved in a volume of 0.1–1 ml of $CHCl_3$:MeOH:water (30:60:8, per vol).

For calibration, HPTLC plates were prepared applying standard solution of the respective APC (100 pmol/μl) in the range of 0.1 to 1.5 nmol in the solvent mixture $CHCl_3$:MeOH:water (30:60:8, per vol). The concentration of standards was determined by the turbidimetric phosphate assay according to Eibl and Lands [14].

For chromatography, the HPTLC plates were prerun with a mobile phase [15] $CHCl_3$/MeOH/triethylamine/water (30:35:34:8, per vol). Samples and calibration solution were streaked automatically to the HPTLC plates with a Linomat IV (CAMAG, Berlin, FRG). The sample volume was selected between 2 and 20 μl. The samples were applied as lines of 5 mm about 10 mm from the lower edge of the plate. The distance to the next sample was 4 mm. The plates were developed in glass tanks with the mobile phase to a distance of about 1 cm from the upper edge, which results in a total distance of 7.5 cm. After drying the plate at 180 °C for 10 min on a heating plate from Desaga (Heidelberg, FRG), visualizing of the lipids occurred by staining with cupric sulfate solution (10%, w/v) in phosphoric acid (8% w/v) [16]. The HPTLC plate was dipped for 15 s in the dye solution and dried on the heating plate by increasing the temperature from 110 to 180 °C within 7 min. The APC spots on the cooled plate were then read quantitatively using a CD 60 TLC scanner from Desaga at 530 nm [17].

Results

A fast and sensitive method for the extraction and quantitative detection of alkylphosphocholines such as HePC, OcPC and ErPC in biological fluids and tissues was developed. Since $ET_{18}OCH_3$ has R_f values similar to

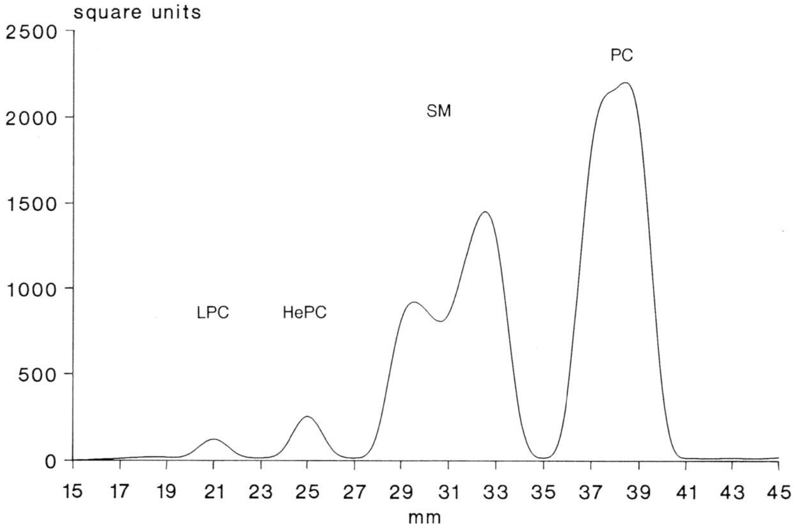

Fig. 1. Densitometric evaluation of a HPTLC plate spiked with human serum phospholipids. The lipids were extracted as described in the 'Methods' section and 5 μl of the redissolved phospholipid extract were applied 10 mm from the lower edge of the prerun HPTLC plate as a line 5 mm long using a Linomat IV. The plate was developed in $CHCl_3$/MeOH/triethylamine/H_2O (30:35:34:8, per vol) up to 90 mm, dried and stained with a 10% copper sulfate solution in 8% phosphoric acid. Subsequently, the lipid spots on the HPTLC plate were evaluated at 530 nm in a CD 60 Densitometer. HePC is clearly separated from lysolecithins (LPC), sphingomyelins (SM) and lecithins (PC) in the range of 23 to 27 mm. This 'window' is also sufficient for the detection of octadecylphosphocholine and 1-O-octadecyl-2-O-methyl-glycero-3-phosphocholine.

OcPC, the analysis is also possible for certain alkyl-lysolecithins and analogs substituted in the 2 position of glycerol. The alkylphosphocholines HePC and OcPC are easily separated from lysolecithins and sphingomyelins [11] with a solvent system [15] $CHCl_3$:MeOH:triethylamine:water (30:35:34:8, per vol) as shown in figure 1. This is not so for EPC, which is only separated from sphingomyelins by the more polar solvent system $CHCl_3$:MeOH:triethylamine:water (30:50:38:8, per vol).

In figure 2, we present a typical set of calibration curves for the four compounds. The absorbance increases linearly with the concentration of the alkylphosphocholines in the range between 0.1 and 1.5 μMol. The regression coefficient at different days for all compounds is 0.995 ± 0.004. The deviations of the calibrated results from the theoretical values are less than 5%.

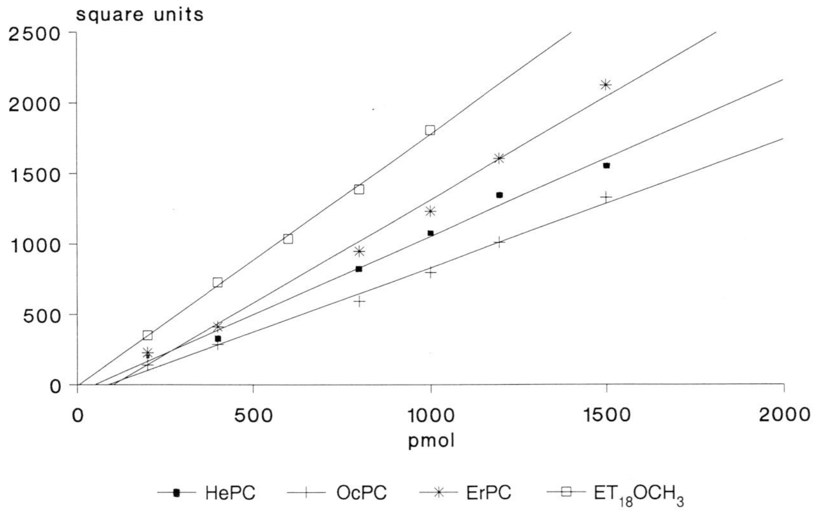

Fig. 2. Calibration curves for HePC, octadecylphosphocholine (OcPC), erucylphosphocholine (ErPC) and 1-O-octadecyl-2-O-methyl-glycero-3-phosphocholine ($ET_{18}OCH_3$) between 0.2 and 1.5 μMol. The square units were evaluated after development and staining of the HPTLC plates with a CD 60 densitometer at 530 nm.

The overall recoveries after extraction and estimation of APCs and $ET_{18}OCH_3$ for the described procedure is summarized in table 1. The recoveries were estimated for tissue concentrations between 5 and 500 nmol/g. In this range of concentrations and for different fluids and tissues the recovery of the compounds is between 90 and 110%. As described above, the composition of the solvent systems for the chromatographic separation of alkylphosphocholines from the membrane phospholipids was carefully selected to obtain an optical clear window in the thin-layer chromatogram. This allowed a sensitive determination of alkylphosphocholines and analogs with the described method. The limit of detection is 50 pmol for alkylphosphocholines and for $ET_{18}OCH_3$.

Taking into account that not more than 20 μl of the redissolved extract should be applied to the HPTLC plate per sample and that a minimum of 100 μl solvent is necessary for redilution of the sample, the lowest APC concentration for precise estimation is about 0.5 nmol/g tissue.

Table 1. Recovery of APC and $ET_{18}OCH_3$ from different tissues of the rat after organic phase extraction and HPTLC

Tissues	HePC	OcPC	ErPC	$ET_{18}OCH_3$
Serum	98 ± 2	98 ± 2	98 ± 2	98 ± 2
Small intestine	100 ± 5	100 ± 5	100 ± 5	100 ± 10
Liver	100 ± 5	100 ± 5	100 ± 5	100 ± 10
Kidney	100 ± 5	100 ± 5	100 ± 5	100 ± 10
Lung	90 ± 2	95 ± 3	98 ± 3	105 ± 10
Brain	85 ± 10	90 ± 5	93 ± 3	95 ± 10
Stomach	80 ± 10	83 ± 5	80 ± 10	70 ± 10
Muscle	90 ± 10	93 ± 5	95 ± 5	100 ± 10

For estimation of the recovery, about 0.5 g of the tissue was spiked with the respective compound to a concentration of 50 nmol/g. Lipids were extracted from the aqueous tissue homogenate with $CHCl_3$/MeOH (2:1, v/v). The percentage recoveries are mean of three independent experiments ±SD.

Discussion

The HPTLC method described here offers a fast and sensitive determination of alkylphosphocholines and $ET_{18}OCH_3$ from samples of biological origin and has significant advantages over the method published earlier [11]. The extraction of APC and $ET_{18}OCH_3$ with $CHCl_3$/MeOH [18, 19] is fast and effective (table 1). The redissolved extracts can be applied to the HPTLC plates without further purification by microcolumns or desalting procedures [11]. Therefore, large numbers of samples can be extracted simultaneously. HPTLC analysis is faster than the HPLC method [10]. Up to 10 samples can be applied beside 7 standards and 3 quality controls on one HPTLC plate.

The APC-containing samples can be stored in a deep freezer (−80 °C) for several months before analysis. There were no significant differences in the APC concentrations in samples measured over a time period of 6 months. Freeze-thawing experiments were also without any influence on the detection of APCs. The $CHCl_3$/MeOH extracts also contain the other major membrane phospholipids quantitatively which can be estimated simultaneously using appropriate standards. The described method is a helpful tool for studies of the pharmacokinetic and pharmacodynamic behaviour of APC and analogues.

References

1 Eibl H, Unger C: Hexadecylphosphocholine: A new and selective antitumor drug. Cancer Treat Rev 1990;17:233–242.

2 Eibl H, Unger C: Biologische Bedeutung der Phospholipide: Entwicklung von Hexadecylphosphocholin – Eine Substanz mit selektiv-antitumoralen Eigenschaften. Stuttgart, Schattauer, 1989.

3 Fleer EAM, Kim DJ, Nagel GA, Eibl H, Unger C: Cytotoxic activity of lysophosphatidylcholine analogues on human lymphoma Raji cells. Onkologie 1990;13:295–300.

4 Stekar J, Hilgard P, Engel J, Schumacher W, Eibl H, Unger C: Hexadecylphosphocholine (D 18506). A new phospholipid with highly selective antitumor activity. J. Cancer Res Clin Oncol 1988;114:39.

5 Vehmeyer K, Kim DJ, Nagel GA, Eibl H, Unger C: Effects of etherlipids on mouse granulocyte-macrophage progenitor cells. Cancer Chemother Pharmacol 1989;24:58–60.

6 Unger C, Damenz W, Fleer EAM, Kim DJ, Breiser A, Hilgard P, Engel J, Nagel GA, Eibl H: Hexadecylphosphocholine, a new ether lipid analogue. Studies on the antineoplastic activity in vitro and in vivo. Acta Oncol 1989;28:213–217.

7 Muschiol C, Berger MR, Schuler B, Scherf HR, Garzon FT, Zeller WJ, Unger C, Eibl H, Schmähl D: Alkylphosphocholines: Toxicity and anticancer properties. Lipids 1987;22:930–934.

8 Hilgard P, Stekar J, Voegeli H, Engel J, Schumacher W, Eibl H, Unger C, Berger MR: Characterization of the antitumor activity of hexadecylphosphocholine (D 18506). Eur J Cancer Res Clin Oncol 1988;28:1457–1461.

9 Unger C, Eibl H: Onkologische Therapie mit Phospholipiden. Stuttgart, Schattauer, 1989.

10 Unger C, Fleer EAM, Damenz W, Hilgard P, Nagel G, Eibl H: Hexadecylphosphocholine: Determination of serum concentrations in rats. J Lipid Mediators 1991;3:71–78.

11 Rustenbeck I, Lenzen S: Quantitation of a new antineoplastic agent, hexadecylphosphocholine. Arch Pharmacol 1990;341 suppl:R111.

12 Eibl H, Woolley P: A general synthetic method for enantiomerically pure ester and ether lyso-phospholipids. Chem Phys Lipids. 1988;47:47–53.

13 Woolley P, Eibl H: Synthesis of enantiomerically pure phospholipids including phosphatidyl serine and phosphatidyl glycerol. Chem Phys Lipids 1988;47:55–62.

14 Eibl H, Lands W: A new sensitive determination of phosphate. Anal Biochem 1969;30:51–57.

15 Touchstone JC, Chen JC, Beaver KM: Improved separation of phospholipids in thin layer chromatography. Lipids 1979;15:61–62.

16 Touchstone JC, Levin SS, Dobbins MF, Carter PJ: Differentiation of saturated and unsaturated phospholipids on thin layer chromatograms. J High Resol Chromatogr Chromatogr Commun 1981;4:423–426.

17 Goppelt M, Resch K: Densitometric quantitation of individual phospholipids from natural sources separated by one-dimensional thin-layer chromatography. Anal Biochem 1984;140:152–156.

18 Bligh EG, Dyer WJ: A rapid method of total lipid extraction and purification. Can J Physiol 1959;37:911–917.

19 Kolarovic L, Fournier NC: A comparison of extraction methods for the isolation of phospholipids from biological sources. Anal Biochem 1986;156:251–256.

Jochem Kötting, Max-Planck-Institut für biophysikalische Chemie,
Am Fassberg, D-W–3400 Göttingen (FRG)

Eibl H, Hilgard P, Unger C (eds): Alkylphosphocholines: New Drugs in Cancer Therapy. Prog Exp Tumor Res. Basel, Karger, 1992, vol 34, pp 12–24

Hexadecylphosphocholine in Liposomal Dispersions

Petra Kaufmann-Kolle[a], *Clemens Unger*[b], *Hansjörg Eibl*[a]

[a]Max-Planck-Institut für biophysikalische Chemie, Abteilung Membranbiophysik, and [b]Medizinische Universitätsklinik, Abteilung Hämatologie/Onkologie, Göttingen, FRG

Introduction

Liposomes are defined as artificial phospholipid vesicles enclosing an aqueous space. In the early 1960s, Bangham et al. [1] first reported that phospholipids in excess of water spontaneously form closed and concentrically orientated multilamellar lipid bilayers. In the last decade, the potential of these artificial vesicles was widely recognized in different disciplines of basic and applied research. For instance, liposomes are examined as drug delivery systems avoiding side effects of drugs. By inclusion into the aqueous space or into the lipid bilayer itself, liposomes may serve as carriers to modify pharmacokinetic and pharmacodynamic parameters of the drugs, as well as to alter metabolic fate or enrichment of drugs in certain tissues [2–7]. Clinical trials with several drugs incorporated in liposomes have been initiated [8–14].

Vesicles differ in size, charge, behavior, and in the number of bilayers, depending on lipid composition and method of preparation [15, 16]. Per definition, small unilamellar vesicles (SUVs) are usually described as particles with a diameter in the range of 20–200 nm, large unilamellar vesicles (LUVs) as particles larger than 200 nm and multilamellar vesicles (MLVs) as particles in the range of 100 to 5,000 nm. Generally, such lamellar structures are only formed by amphiphiles like lecithin with a ratio of about one for the surface requirement of the polar part divided by the surface requirement of the apolar part of the molecule. In the case of lysolecithins this ratio is >1. Therefore, lysolecithins form micelles when dissolved in water in concentrations higher than the critical micellar concentration (CMC). Micelles do not have an internal space, are instable to dilution and thus they strongly differ from liposomes [17].

Hexadecylphosphocholine (HePC) and other alkylphosphocholines are lysolecithin-like structures with strong antineoplastic activity [18]. In water they form micelles instead of liposomes. This behavior can be altered by addition of cholesterol which obviously will increase the surface requirement of the apolar region in lipid mixtures. Now, liposomes can be obtained. This might become very important, if for instance HePC or other lipids should be applied intraveneously instead of topically or orally. In this report, we describe the preparation of liposomes containing HePC and some of their properties.

Materials and Methods

Materials

HePC was synthesized in our laboratory as described in this volume by H. Eibl and J. Engel. The synthesis of the other phospholipids was described earlier [19–22]: 1,2-O,O-dihexadecyl-*rac*-glycero-3-phospho-N,N-dimethyl-N-(ethyltrimethyl-ammonium chloride)-ethanolamine (HH-G-3-P-N^+N^+); 1,2-dimyristoyl-3-(N,N,N-trimethyl)-ammonium chloride; (MM-G-3-N^+) 1,2-dipalmitoyl-*sn*-glycero-3-phosphocholine (PP-G-3-PC); 1,2-dipalmitoyl-*sn*-glycero-3-phosphoglycerol (PP-G-3-PG); 1-lauroyl-*rac*-glycerol (1-L-G). Cholesterol (Chol) was purchased from Fluka Chemie AG (Buchs, Switzerland) and stearylamine (SA) from Sigma Chemie GmbH (Deideshofen, FRG).

Preparation of Liposomes

Lipids suspensions, 50 m*M* in 154 m*M* NaCl, were mechanically dispersed in a rotary evaporator for 12 h at temperatures 5 °C above the phase transition temperature (T_M) of the lipid with the highest T_M. This treatment resulted in a homogeneous lipid suspension. From this, small unilamellar vesicles were prepared by the application of three cycles at 20,000 psi using the French pressure cell (Aminco, J5-598A, Silver Spring, Md., USA). After 30 min centrifugation at 30,000 *g* (J2-21M/E, Beckman Instruments GmbH, München, FRG) particles and multilamellar vesicles were removed by precipitation. The supernatant contained SUVs.

Photon Correlation Spectroscopy (PCS)

The vesicle size distribution and the mean hydrodynamic diameter were determined by dynamic light scattering analysis which is also referred to as quasi-elastic light scattering or photon correlation spectroscopy. The scattering angle was 90°, using a Nicomb 370 submicron particle sizer (Pacific Scientific, Silver Spring, Md., USA), equipped with a Helium Neon Laser (632.8 nm). Nicomp 370 employs a digital autocorrelator to analyze the fluctuations in scattered light intensity generated by the diffusion of SUVs in the dispersions. The measured diffusion coefficient is used in the Stokes-Einstein relation to obtain the average radius and, hence, the mean hydrodynamic diameter of liposomes. Our dispersions showed a bimodal distribution, a so-called Nicomb relation with a fit error <2. For standards we used latex suspensions (nanosphere size standards, Distrilab, Leusden, The Netherlands).

Microscopy

For negative staining, 10 μl of the SUV dispersion were dropped on copper grids, which were covered with thin layers of carbon supported by a film of pioloform (FN50, Wacker-Chemie, Hannover, FRG). After washing, the adsorbed SUVs were stained with a solution of 1% (g/v) uranyl acetate in distilled water. Transmission electron micrographs (TEM) were recorded on a

Jeol 100B instrument operating at 100 kV under normal beam illuminating conditions. The instrumental magnifications were within the range of 18,000–24.000 ×. For further details see Kaufmann-Kolle [15]. For freeze fracture, 10 μl of the well-concentrated dispersions were flash-frozen in liquid propane, then transferred to liquid nitrogen and fractured under vacuum in a BAF 400 (Balzers Hochvakuum, Balzers, Liechtenstein). These preparations were shadowed using platinum/carbon at a 45° angle and afterwards carbon at a 90° angle, all at −105 °C under vacuum. Replicas were cleaned with a mixture of choloroform/methanol (1/1) and observed by transmission electron microscopy (TEM) as described above. For light microscopy (LM) a Zeiss microscope with a Plan-Neofluar 25/0.6 objective (Zeiss, Göttingen, FRG) was used.

Analysis of Lipid Content

The ratio of the different lipid compounds before and after preparation with the French press was examined by HPTLC as described by Kötting et al. [23]. For determination of the lipid content we used the turbidimetric phosphate assay according to Eibl and Lands [24].

Fluorescence Spectroscopy

Release of 6-carboxyfluorescein (CF, Eastman Kodak, Rochester, N.Y., USA) was assayed according to Weinstein et al. [25]. The concentration of CF within the internal aqueous space of the SUV was assumed to be 200 m*M*, which was the solution concentration used to suspend the lipids. At this concentration CF is self-quenching. After preparation, the not-included CF was separated from the CF-filled SUVs by gel filtration (Sephadex G-75, LKB-Pharmacia, Freiburg i.Br., FRG). The entrapment efficiency and the release of CF from liposomes were determined using a Hitachi F-4000 fluorescence spectrophotometer (wavelength: excitation 491 nm, emission 515 nm; Colora Messtechnik, Hannover, FRG). Serum was gained from blood of human volunteers. It was mixed with K^+EDTA in monovettes (Saarstedt, Salzgitter, FRG) and centrifuged for 5 min at 600 *g*. To measure the total amount of entrapped CF, SUVs were disrupted with Triton X-100 (Boehringer, FRG) in a final concentration of 1%. SUVs in different concentrations were incubated at 37 °C. At different time intervals, aliquots were taken and sufficiently diluted. The intensity of the fluorescence was proportional to the concentration of CF in the range of 20–300 n*M*.

Hemolytic Activity

Hemolysis of HePC liposomes was determined as described by Kötting et al. [26].

Results

Development of Stable Liposome Dispersions

As shown in table 1, the ability of SUV formation and the stability of the resulting SUV dispersion strongly depends on several facts. For instance, in water above the CMC HePC as an amphiphilic single-chain alkylphosphocholine does form micelles. This results in an optical clear solution. This was also observed for mixtures of HePC with 1-lauryl-*rac*-glycerol. However if the molar amount of HePC in lipid mixtures does not exceed 30%, SUVs can be prepared by treatment with the French

Table 1. Influence of the lipid composition on the stability of colloidal systems in water

Lipids	Molar ratio	Result after preparation	Result after 3 days of storage at 20 °C
HePC		micelles	micelles
HePC/1-L-G	1/1	micelles	micelles
HePC/PP-G-3-PC	2/1	micelles	precipitation
HePC/PP-G-3-PC	1/1	micelles	precipitation
HePC/PP-G-3-PC	1/2	SUVs	precipitation
HePC/Chol/PP-G-3-PC	1/1/1	SUVs	milky suspension
HePC/Chol/SA	4/5/1	SUVs	milky suspension
HePC/Chol/HH-G-3-$P^-N^+N^+$	4/5/1	SUVs	milky suspension
HePC/Chol/MM-G-3-N^+	4/5/1	SUVs	milky suspension
HePC/Chol	1/1	SUVs	milky suspension
HePC/Chol/PP-G-3-PG	1/1/1	SUVs	SUVs
HePC/Chol/PP-G-3-PG	1/2/1	SUVs	SUVs
HePC/Chol/PP-G-3-PG	2/4/1	SUVs	SUVs
HePC/Chol/PP-G-3-PG	4/5/1	SUVs	SUVs
HePC/Chol/PP-G-3-PG	5/5/1	SUVs	SUVs

After treatment with the French pressure cell the resulting products, which originally had a concentration of 50 m*M* in a solution of 154 m*M* NaCl, were centrifuged. Supernatant was judged by LM and TEM, respectively. SUVs = small unilamellar vesicles in an opalescent shining dispersion; micelles = clear solution.

pressure cell from suspensions which contain HePC and phosphatidylcholines, as for example PP-G-3-PC. Unfortunately, the resulting opalescent dispersions of SUVs are not very stable and precipitate after storage within 3 days. Suspensions which consist of HePC and cholesterol in a molar ratio of 1/1 and mixtures of HePC, PP-G-3-PC and cholesterol in a molar ratio of 1/1/1 show a similar behavior. The treatment of these suspensions with the French pressure cell followed by centrifugation results in formation of opalescent shining dispersions, which mainly contain SUVs. But again, these dispersions are instable and change into milky suspensions within 3 days. By microscopy we can observe large multilamellar structures and precipitates which are probably cholesterol particles. This instability of liposome dispersions can be avoided by the addition of negatively charged phospholipids such as phosphatidylglycerols. In this respect, PP-G-3-PG has favorable properties. As shown in table 3 the molar ratio of the lipid components is only of interest with respect to the molar amount of cholesterol which should be equal or exceed the molar amount of HePC. The ratio of phosphatidylglycerol can vary from 5 to 30%. These liposome dispersions are stable at 4 °C for time periods of at least 1 year. Finally, it is interesting to report that stable dispersions

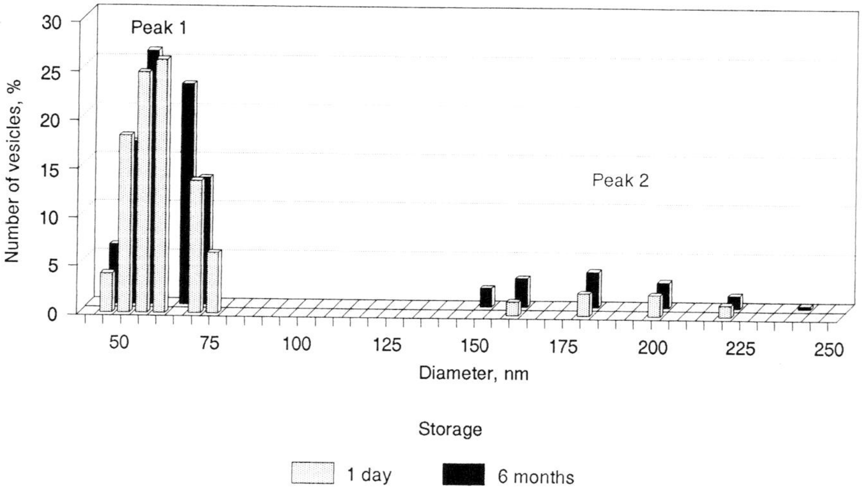

Fig. 1. Size distribution of SUVs containing HePC/Chol/PP-G-3-PG (4/5/1). Liposome size distribution was determined by PCS directly after preparation and after storage for 6 months at 4 °C. Size is expressed as mean hydrodynamic diameter: peak 1, 60 nm (1 day), 59 nm (6 months); peak 2, 185 nm (1 day), 179 nm (6 months).

of SUVs were not obtained if positively charged lipids like HH-G-3 $P^{-}N^{+}N^{+}$, MM-G-3-N^{+}, or stearylamine were used instead for negatively charged phosphatidylglycerols (table 1).

Evaluation of Size, Number of Bilayers and Lipid Content

After preparation with the French pressure cell and after centrifugation, usually a lipid recovery of 75% was found in the supernatant for SUV dispersions consisting of HePC/Chol/PP-G-3-PG (4/5/1). As shown by HPTLC the ratios of the different lipid compounds before and after preparation with the French press were equal. The SUV size distribution of the dispersions was measured by PCS. As shown in figure 1, the population of SUVs in the dispersions was quite homogeneous. The distribution is bimodal but most vesicles (>90%) have a hydrodynamic diameter between 45 and 75 nm with a mean hydrodynamic diameter of 69 nm. Unilamellarity of the SUVs in these dispersions was proved by TEM. Observed vesicles were small, round and homogenous, each consisting of only one bilayer (fig. 2a, negative staining; fig. 2b, freeze fracture). Their diameters were in the same range as measured by PCS.

Hemolytic Properties of SUV Dispersions

HePC in micellar solution causes 50% spontaneous hemolysis in the described test at a concentration of 5 μM. The liposomal dispersions of

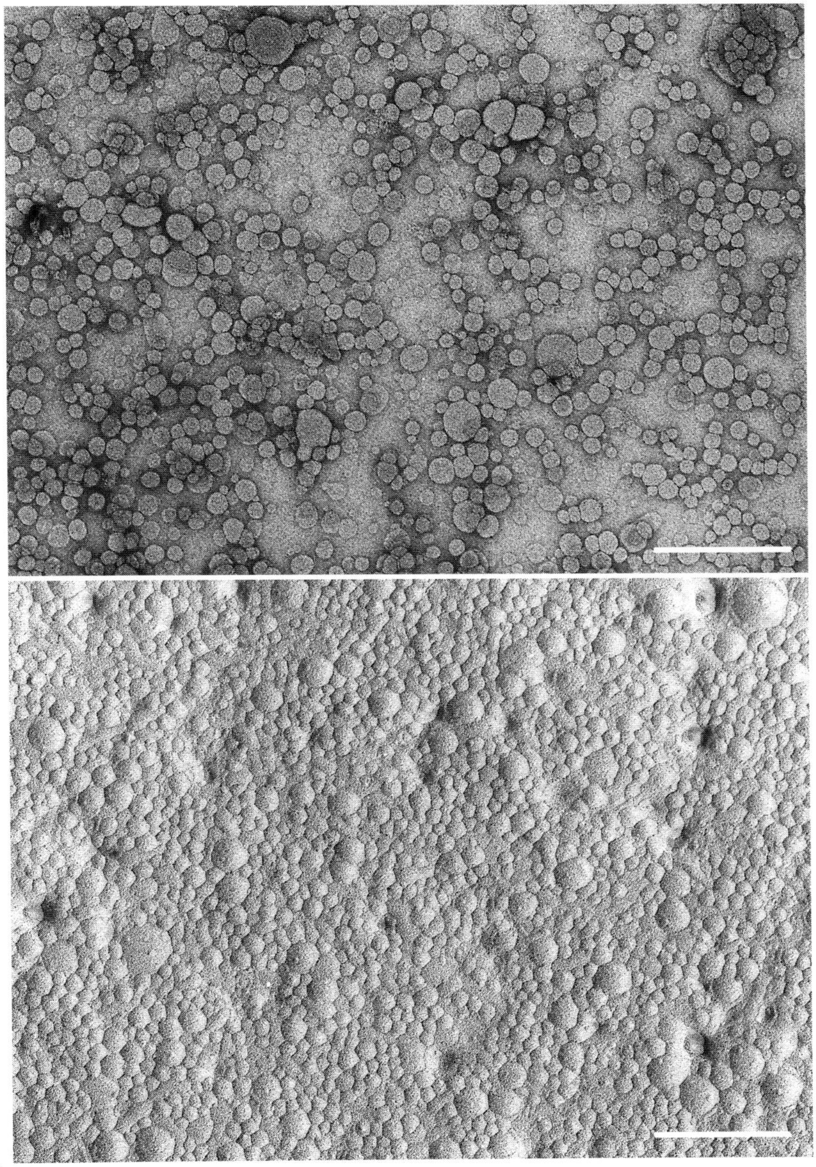

Fig. 2. Transmission electron micrographs: negative staining (2*a*); freeze fracture (*b*). Bar = 200 nm.

Table 2. Hemolytic activity of SUV dispersions depending on different molar ratios of HePC, Chol and PP-G-3-PG as well as on storage conditions

No.	Molar ratio	Concentration of liposomal HePC (μM), which causes 50% spontaneous hemolysis				LM
		2 weeks	3–6 weeks	15–30 weeks	50 weeks	
1	4/5/1	110	123	150	126	transparent
2	4/5/1	115	n.d.	136	108	transparent
3	4/5/1	55	n.d.	28	31	transparent
4	4/5/1*	n.d.	62	58	73	transparent
5	4/5/1*	n.d.	45	28	26	milky suspension
6	5/5/1	20	n.d.	15	15	transparent
7	9/9/1	22	n.d.	16	15	transparent
8	1/2/1	>1,000	n.d.	n.d.	>1,000	transparent
9	1/2/1*	>1,000	n.d.	n.d.	>1,000	milky suspension

The concentrations of HePC in the SUV dispersions, which cause 50% spontaneous hemolysis are given in μM. Under the same test conditions, HePC in a micellar solution without other lipids causes 50% spontaneous hemolysis at a concentration of 5 μM. All SUV dispersions were stored at 4 °C, except for those marked with an asterisk which were kept at 20 °C. LM gives the optical result after storage of 45–50 weeks. n.d. = Not done; transparent = opalescent shining SUV dispersion without visible sedimentation; milky suspension = agglomeration and sedimentation has taken place.

HePC are less hemolytic which is indicated by the higher HePC concentration needed to induce the same hemolytic effect. Obviously, the hemolytic activity of HePC depends on the molar ratio of HePC to cholesterol as shown in table 2. If the molar amount of cholesterol exceeds the molar amount of HePC, the protection of erthrocytes against hemolysis is high and at least better than a factor of ten. However, if the ratio of molar amounts of HePC to cholesterol becomes 1 or >1, the hemolytic activity of the SUV dispersion increases.

Stability of Liposomes during Storage

The hemolytic activity of HePC in liposomal dispersions is influenced by storage conditions (table 2). There is no clear relationship between the quality of SUV dispersions and the hemolytic potency. For example, dispersions consisting of HePC/Chol/PP-G-3-PG (4/5/1) which were stored at 4 °C and, according to TEM and LM, were in good condition, showed differences in hemolytic activities (table 2, e.g. No. 1, 2 and 3). Surprisingly, in some cases hemolytic activity of stable opalescent shining dispersions do not differ very much from the hemolytic activity of milky suspensions containing the same lipids but which have been stored at 20 °C

Table 3. Long-term stability of CF-filled SUVs, which consist of HePC/Chol/PP-G-3-PG (4/5/1) and are stored at 4 °C in a solution of 154 m*M* NaCl

	Period			
	1 day	1 month	2 months	3 months
Latency, %	82	81	80	80

The amount of included CF is given as percentage of total.

instead of 4 °C (table 2, comparing No. 3 and 5). As demonstrated by No. 6, 7, 8 and 9 (table 2), hemolytic activity of HePC solely depends on the ratio of HePC to cholesterol. Ratios <0.5 result in structures with extremely low hemolytic activity.

Important information, about the mean size and about the alterations in the size distribution of liposome dispersions can be obtained by photon correlation spectroscopy (PCS). Immediately after preparation, a SUV dispersion of HePC/Chol/PP-G-3-PG (4/5/1) shows a bimodal size distribution. In such a dispersion 92% of the SUVs have a mean hydrodynamic diameter of 60 nm and 8% of the SUVs show a mean hydrodynamic diameter of 185 nm (fig. 1). After 6 months of storage at 4 °C there are no dramatic changes in the mean diameter of the dispersion. As also shown in table 2, after storage at 4 °C for nearly 1 year, no sedimentation or agglomeration can be detected by TEM or LM. When the same dispersions were stored at 20 °C, some of them but not all show sedimentation (table 2). Finally, we have shown by incorporation of 6-carboxyfluorescein (CF) in liposomes consisting of HePC/Chol/PP-G-3-PG (4/5/1) that the integrity of the vesicle membranes is given and no destruction has taken place. As shown in table 3, after storage over a period of 3 months at 4 °C, these liposomes are stable based on the observation that almost no leakage of CF can be observed.

Stability of Liposomes in Human Serum

The release of CF from SUVs consisting of HePC/Chol/PP-G-3-PG (4/5/1) is dependent on the concentration ratio of liposomes to serum (fig. 3). At concentrations above 2.6 μmol liposomal HePC per ml fresh human serum there is nearly no CF leakage. Whereas at concentrations beneath 2.6 μmol, the efflux of CF out of the SUVs is biphasic with an initial period of rapid release (in the first 5 min) and a following period of nearly no leakage. At point zero, all dispersions contain 8% of free CF. When incubated at 37 °C in a solution of 154 m*M* NaCl instead of human serum, the same liposomes show no leakage of CF.

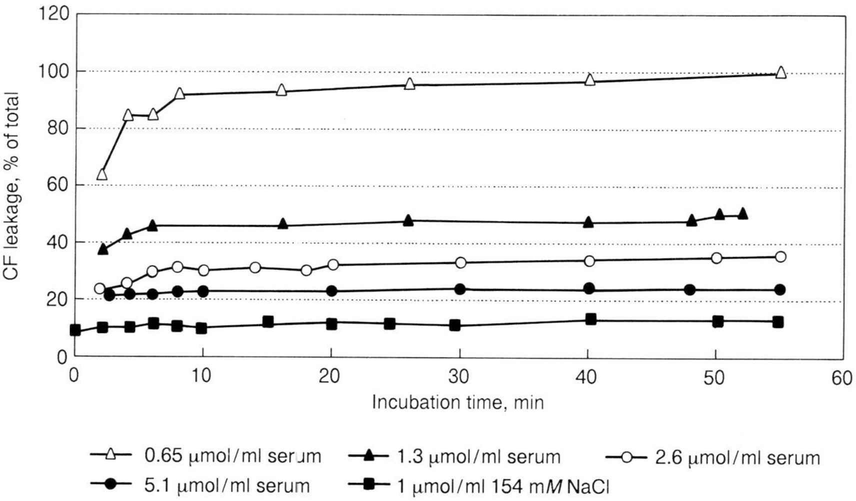

Fig. 3. Incubation of SUVs in human serum or buffer. Time course of CF release from SUV dispersions consisting of HePC/Chol/PP-G-3-PG (4/5/1) during incubation at 37 °C. Concentrations are given as HePC content (μmol) of the liposome dispersion. The leakage is expressed as percentage of total CF which was determined after disrupting the liposomes with Triton X-100.

Discussion and Conclusion

Concerning the molecular shape, HePC is closely related to lysophosphatidylcholines. The surface requirement of the polar part is larger than that of the apolar part which results in a ratio polar to apolar >1. Therefore, HePC typically forms micelles in aqueous solutions above the CMC. Because of hemolysis and necrosis at the injection site, HePC cannot be applied intravenously as a micellar solution [unpubl. data]. Therefore, it was the aim of this study to search for a dispersion which allows i.v. injection of HePC.

In natural membranes lysophosphatidylcholines are only found in small amounts, mainly as intermediates of the bioconversion of phospholipids. Normally, lysophospholipids in higher concentrations have destructive effects on natural and artificial membranes, because of their lytic activity [27–29]. Cholesterol reduces the lytic properties. In addition, liposomes can be formed from lysolecithins [30] or analogues [15]. This principle is also verified in the case of niosomes. These are vesicles which consist of cholesterol and synthetic non-ionic single-chain surfactants instead of phospholipids [31, 32]. Niosomes seem to be very stable [33] and are predominately used in cosmetics.

In this study, the toxicity of HePC is reduced by the combination with cholesterol. In addition, small amounts of negatively charged phospholipids such as phosphatidylglycerols are used to stabilize the SUV dispersions at 4 °C for at least 1 year. One of our preferred lipid combinations in this respect consists of HePC/Chol/PP-G-3-PG in molar ratios of 4/5/1.

For estimation of size, size distribution, numbers of bilayers and stability different methods have been discussed in this paper. For the evaluation of liposome dispersions, it is advisable to combine several methods, namely PCS, TEM, LM and fluorescence spectroscopy. The determination of hemolytic activity is not sufficient for the characterization of these liposomes since even dispersions with enhanced sedimentations may not show hemolytic activity. Independent of size, structure and charge and irrespective of the organization of the lipids in SUVs, LUVs, MLVs or other lamellar structures, the only important parameter for the hemolytic potency of such preparations is the ratio of cholesterol to HePC. This observation is in good agreement with other reports [15, 30]. A molar ratio of HePC to cholesterol <1 in liposomes will reduce the hemolytic activity by a factor larger than five.

Liposomes can interact with cells and biological tissues in different ways. For instance, fusion, adsorption, phagocytosis, contact release and intermembrane transfer have been discussed and demonstrated in different test systems [15, 16]. For medical application, the stability of liposomes in human serum is important. We have studied CF-filled liposomes consisting of HePC/Chol/PP-G-3-PG. The observed dye release may be a result of the destruction of the vesicles induced by a lipid transfer to lipoproteins. Such a disintegrating lipid transfer in vitro and in vivo from liposomes to lipoproteins with following rapid redistribution among the different classes of lipoproteins is also described for SUVs, prepared from egg lecithin without cholesterol [34]. Besides other lipids the incorporation of cholesterol (30% or more) into membranes of SUVs generally inhibits this disintegration by lipoproteins in vitro and in vivo [15, 34]. Such cholesterol containing SUVs remain intact with complete retention of their internal contents and show longer half life time in circulation [15].

Hence, one would expect a similar biological stability, as described above, for liposomes consisting of HePC/Chol/PP-G-3-PG which contain 50% cholesterol. However, these liposomes are not very stable in human serum. We do not expect that this destruction of vesicles is induced by the negatively charged PP-G-3-PG. As we know from other studies, SUVs which consist of PP-G-3-PC/Chol/PP-G-3-PG (5/4/1) show a long half-life time in vivo [15]. Since HePC shows a strong affinity to lipoproteins [26], most probably this interaction with components of the serum is responsible for the rapid destruction of liposomes containing HePC.

Very often liposomes are used for site-selective delivery of drugs [2–16]. In the case of SUVs consisting of HePC/Chol/PP-G-3-PG this type of targeting is not necessary. The only aim of this study was to protect normal tissue against the lytic properties of HePC and to reduce its local toxicity but maintain the antineoplastic activity. This is indeed possible, since the first in vivo tests in different laboratories with healthy and tumor-bearing rats demonstrated that liposomal HePC can be injected intraveneously without necrosis at the injection site and without any loss of antitumoral efficacy [personal commun., M.R. Berger, DKFZ, Heidelberg, and J. Stekar, ASTA PHARMA, Bielefeld, FRG]. In addition, the well-tolerated dose of liposomal HePC in oral application is much higher than the dose of HePC applied in micellar solution [unpublished observations].

From the obtained results it can be concluded that liposomes are a promising vehicle for delivery of HePC in vivo. Small unilamellar liposomes consisting of HePC/Chol/PP-G-3-PG in the molar ratio 4/5/1 can be stored at 4 °C for at least 6 months and used for intravenous injection.

References

1 Bangham AD, Standish MM, Watkins JC: Diffusion of univalent ions across the lamellae of swollen phospholipids. J Mol Biol 1965;13:238–252.

2 Gregoriadis G, Poste G: Targeting of drugs: Anatomical and physiological considerations; in Gregoriadis G, Poste G (eds): Targeting of Drugs. Proceedings of the NATO Advanced Institute on Targeting of Drugs. New York, Plenum Press, 1988, pp 153–176.

3 Gregoriadis G: Fate of injected liposomes: Observations on entrapped solute, retention vesicle clearance and tissue distribution in vivo; in Gregoriadis G (ed): Liposomes as Drug Carriers: Recent Trends and Progress. New York, Wiley, 1988, pp 3–18.

4 Gregoriadis G: Liposomes as drug delivery systems: Optimization studies. Adv Exp Med Biol 1988;238:151–159.

5 Gabizon A: Liposomes as drug delivery system in cancer chemotherapy; in Roerdink FHD, Kroon AM (eds): Drug Carrier Systems. New York, Wiley, 1989, pp 185–211.

6 Ostro MJ, Cullis PR: Use of liposomes as injectable drug delivery systems. Am J Hosp Pharm 1989;46:1576–1587.

7 Szoka FC: The future of liposomal drug delivery. Biotechnol Appl Biochem 1990; 12:496–500.

8 Sculier JP, Delroix C, Brassinne C, Laduron C, Hollaert C, Coune A: Pharmacokinetiks of Amphotericin B in patients receiving repeated intravenous high doses of amphotericin B entrapped into sonicated liposomes. J Liposome Res 1989;2:151–166.

9 Creaven PJ, Cowens JW, Ginsberg R, Ostra M, Browman G: Clinical studies with liposomal doxorubicin. J Liposome Res 1990;4:481–490.

10 Gabizon A, Amselm S, Goren D, Druckmann S, Fromer I, Chisin R, Peretz T, Sulkes A, Barenholz Y: Preclinical and clinical experience with a doxorubicin liposome preparation. J Liposome Res 1990;4:491–502.

11 Lopez-Berestein G: Therapeutic activity of liposomal amphotericin B in systemic mycoses. J Liposome Res 1990;2:151–166.
12 Presant CA, Ksionski G, Crossley R: In-labelled liposomes for tumor imaging: Clinical results of the liposome imaging. J Liposome Res 1990;4:431–436.
13 Perez-Soler R, Knokhar AR, Lautersztain J, Albaker S, Francis K, Maciaskiger D, Lopez-Berestein G: Clinical development of liposomal platinum. J Liposome Res 1990;4:437–449.
14 Urba WJ, Hartmann LC, Longo DL, Steis RG, Smith JW, Kedar I, Creekmore S, Sznol M, Conlon K, Kopp WC, Huber C, Herold M, Alvord WG, Snow S, Clark JW: Phase 1 and immunomodulatory story of a muramyl peptide, muramyl tripeptide phosphatidyletanolamine. Cancer Res 1990;50:2979–2986.
15 Kaufmann-Kolle P: Liposomen aus synthetischen Phospholipiden: Struktur, Stabilität. und biologisches Verhalten; Diss, Technische Universität, Branschweig, 1986.
16 New RRC, Black CDV, Parker RJ, Puri A, Scherphof: Liposomes in biological systems; in New RRC (ed): Liposomes, a Practical Approach. The Practical Approach Series. Oxford, IRL Press at Oxford University Press, 1990, pp 221–252.
17 Eibl H: Phospholipide als funktionelle Bausteine biologischer Membranen. Angew Chem 1984;96:247–262.
18 Eibl H, Unger C: Hexadecylphosphocholine, a new and selective antitumor drug. Cancer Treat Rev 1990;17:233–242.
19 Eibl H, Woolley P: Synthesis of enantiomerically pure glyceryl esters and ethers. 1. Methods employing the precursor 1,2-isopropylidene-*sn*-glycerol. Chem Phys Lipids 1986;41:53–63.
20 Eibl H, Woolley P: Synthesis of enantiomerically pure glyceryl esters and ethers. 2. Methods employing the precursor 1,2-isopropylidene-*D*-mannitol. Chem Phys Lipids 1988;47:47–53.
21 Woolley P, Eibl H: Synthesis of enantiomerically pure phospholipids including phosphatidylserine and phosphatidylglycerol. Chem Phys Lipids 1988;47:55–62.
22 Eibl H, Woolley P: A general synthetic method for enantiomerically pure ester and ether lysophospholipids. Chem Phys Lipids 1988;47:63–68.
23 Kötting J, Marschner N, Unger C, Eibl H: Determination of alkylphosphocholines in biological fluids and tissues. Prog Exp Tumor Res. Basel, Karger, 1992, vol. 34, pp. 6–11.
24 Eibl H, Lands WEM: A new sensitive determination of phosphate. Anal Biochem 1969;30:51–57.
25 Weinstein JN, Yoshikami S, Henkart P, Blumenthal R, Hagins WA: Liposome-cell interaction: Transfer or intracellular release of a trapped fluorescent marker. Science 1977;195:489–492.
26 Kötting J, Marschner N, Neumüller W, Unger C, Eibl H: Hexadecylphosphocholine and octadecyl-methyl-glycero-3-phosphocholine: A comparison of hemolytic activity, serum binding and tissue distribution: Prog Exp Tumor Res. Basel, Karger, 1992, vol 34, pp. 131–142.
27 Reman FC, Demel RA, Degier J, Van Deenen LLM, Eibl H, Westphal O: Studies of the lysis of red cells and biomolecular lipid leaflets by synthetic lysolecithins, lecithins and structural analogs. Chem. Phys. Lipids 1969;3:221–233.
28 Harlos K, Vaz WLC, Kovatchev S: Effect of desoxylysolecithins on dimyristoyllecithin vesicles. FEBS Lett 1977;77:7–10.
29 Mandersloot JG, Reman FC, Van Deenen LLM, De Gier J: Barrier properties of lecithin/lysolecithin mixtures. Biochim Biophys Acta 1975;382:22–26.
30 Kitagawa T, Inoue K, Nojima S: Properties of liposomal membranes containing lysolecithin. J Biochem 1976;79:1123–1133.

31 Handjani-Vila RM, Ribier A, Rondot B, Vanlerberghe G: Int J Cosm Sci 1979;1:303.
32 Okahata Y, Tanamachi S, Nagai M, Kunitake T: J Colloid Interface Sci 1981;82:401.
33 Baillie AJ: Niosomes: A putative drug carrier system; in Gregoriadis G, Poste G (eds): Targeting of Drugs. Proceedings of the NATO Advanced Institute on Targeting of Drugs. New York, Plenum Press, 1988, pp 143–151.
34 Scherphof G, Roedrink F, Waite M, Parks J: Disintegration of phosphatidylcholine liposomes in plasma as a result of interaction with high-density lipoproteins. Biochim Biophys Acta 1978;542:296–307.

Petra Kaufmann-Kolle, Max-Planck-Institut für biophysikalische Chemie, Abteilung Membranbiophysik, Am Fassberg, D-W–3400 Göttingen (FRG)

Eibl H, Hilgard P, Unger C (eds): Alkylphosphocholines: New Drugs in Cancer Therapy. Prog Exp Tumor Res. Basel, Karger, 1992, vol 34, pp 25–32

Antitumoral Activity of Alkylphosphocholines and Analogues in Human Leukemia Cell Lines

Clemens Unger[a], *Eduard A.M. Fleer*[a], *Jochem Kötting*[b], *Waldemar Neumüller*[b], *Hansjörg Eibl*[b]

[a]Medizinische Universitätsklinik, Abteilung Hämatologie/Onkologie, and [b]Max-Planck-Institut für biophysikalische Chemie, Göttingen, FRG

Introduction

Alkylphosphocholines [1] and (alkyl)-lysophospholipids [2] are substances with a wide spectrum of biological properties such as cytotoxicity against tumor cells in vitro and in vivo [1–6], induction of differentiation in malignant cells [7], activation of cytotoxic macrophages [8] and inhibition of neoplastic cell invasion in normal tissues [9]. A comprehensive overview is given in the Proceedings of the First International Symposium on 'Ether Lipids in Oncology' 1987 [10]. In recent years we have intensively studied the structure-activity relationships of (alkyl)-lysophospholipids [2, 6, 11–14]. From the results obtained we learned that cytotoxic activity depends on certain requirements [15]. Firstly, the compound should be metabolically stable and maintain a lysolecithin-like structure. Therefore, reacylation in the sn-2-position of glycerol as well as alkyl cleavage of the alkyl residue in the sn-1 position must be prevented. Secondly, phosphocholine is essential for antineoplastic activity. Knowledge about these structural requirements for antineoplastic activity was used to prepare more simple molecules, formed by esterification of long chain alkanols with phosphocholine. We also investigated analogues where the phosphorus-nitrogen distance was elongated from two to six methylene groups. Here we report on the in vitro antineoplastic activity of five alkylphosphocholines and analogues in comparison to (alkyl)-lysophospholipids.

Materials and Methods

Materials

Alkylphosphocholines and analogues were prepared according to general procedures for the conversion of alcohols to their respective phosphocholines as described by

Eibl and Woolley [18]. 1-Hexadecylmercapto-2-methoxymethyl-rac-glycero-3-phosphocholine (BM 41.440) was a gift from Boehringer Mannheim, FRG. 1-0-octadecyl-2-0-methyl-rac-glycero-3-phosphocholine (Oc-Me-G-3-PC) was synthesized as described elsewhere [19, 20]. Purity of the compounds before starting the experiments was checked by thin-layer chromatography and was found to be greater than 98%.

Cell Lines

Four human leukemia cell lines were used in the study: Raji (B lymphoblastic lymphoma); HL 60 (acute promyelocytic leukemia); K 562 (chronic myelogenous erythroleukemia), and U 937 (promyelocytic leukemia). All cell lines were grown in RPMI medium (Biochrom, Berlin, FRG), supplemented with 10% fetal calf serum, penicillin (105 IU/l), and streptomycin (128 mg/l).

Determination of Cell Viability and Proliferation

The cytotoxic effects of the lipids were determined by trypan blue dye exclusion at various incubation times. The proliferative capacity of leukemia cells was determined by (^{3}H) thymidine uptake. Three- to sixfold replicates of 10^4 tumor cells/well were incubated with the compounds at concentrations between 1 and 100 μM for 48 h at 37 °C in 96-well microculture trays (Greiner, Nürtingen, FRG). A 6-hour pulse with (^{3}H)thymidine (spec. actitivy 10 Ci/mmol; Amersham Buchler) at 0.5 μCi/well was performed at the end of the culture period. Cells were harvested with a semiautomated cell harvester (PDH, Cambridge Technology Inc., Cambridge, Mass., USA). Filter dishes were air dried, and radioactivity was measured by liquid scintillation counting.

Erythrocyte Hemolysis

Venous blood was collected from healthy young volunteers, mixed with 100 IU heparin/ml blood and subsequently diluted with 9 parts of 0.9% NaCl. After centrifugation (2,600 *g*, 10 min, 20 °C), the supernatant was discarded and the sediment was washed 3 times with 0.9% NaCl. Final hematocrit was adjusted to 45%. Measurement of hemolysis of the erythrocytes was carried out spectrophotometrically according to the method described by Reman et al. [21].

Results

The antineoplastic effect of alkylphosphocholines and derivatives was tested on four different leukemia cell lines in culture. Figure 1 shows the structure of the compounds used in the study. The first three substances differ only in the apolar chain. Compound 1 is hexadexylphosphocholine (HePC) with a chain length of 16 carbons. Compound 2, octadecylphosphocholine (OcPC), has a chain length of 18 carbons and compound 3, oleylphosphocholine (OlPC), has an 18-carbon chain with one cis-unsaturated bond between position 9 and 10. Compound 4, octadecyl-phospho-(N,N,N-trimethylammonio)-hexanol ($OcPC_6$) and compound 5, oleyl-phospho-(N, N, N-trimethylammonio)-hexanol ($OlPC_6$), show a modification in the phosphorus-nitrogen distance. Instead of two methylene groups in choline, the insertion of six methylene groups leads to an increased distance between phosphorous and nitrogen.

1 $CH_3-(CH_2)_{15}-O-PO_3^--(CH_2)_2-N^+(CH_3)_3$

2 $CH_3-(CH_2)_{17}-O-PO_3^--(CH_2)_2-N^+(CH_3)_3$

3 $CH_3-(CH_2)_7-CH{=}CH-(CH_2)_8-O-PO_3^--(CH_2)_2-N^+(CH_3)_3$

4 $CH_3-(CH_2)_{17}-O-PO_3^--(CH_2)_6-N^+(CH_3)_3$

5 $CH_3-(CH_2)_7-CH{=}CH-(CH_2)_8-O-PO_3^--(CH_2)_6-N^+(CH_3)_3$

Fig. 1. Chemical structures of alkylphosphocholines and analogues. 1 = hexadecylphosphocholine (HePC); 2 = octadecylphosphocholine (OcPC); 3 = oleylphosphocholine (OlPC); 4 = octadecylphospho-(N,N,N-trimethylammonio)-hexanol ($OcPC_6$); 5 = oleylphospho-(N,N,N-trimethylammonio)-hexanol ($OlPC_6$).

The antiproliferative activity of the different substances was determined by thymidine incorporation, whereas the trypan blue dye exclusion assay was chosen to measure cell death. Both, the 50% growth inhibition dosage (ID_{50}) in the (^3H)thymidine assay and the half-lethal dosage (LD_{50}) in the trypan blue dye assay were calculated after 48 h of cell incubation, when cell growth was in a logarithmic phase. Table 1 shows the ID_{50} values obtained in leukemia cell lines after 48 h of drug exposure. The most sensitive cell line was U 937 followed by HL 60 and Raji, whereas K 562 was relatively resistant against these compounds. The most effective compound, OcPC, showed ID_{50} values between 1.1 and 15.5 μM. A reduction of the alkyl chain by two carbons, as in HePC, diminished the antiproliferative effect by a factor of >3. A pronounced reduction in the antineoplastic activity was also observed with the introduction of an unsaturated bond in the long alkyl residue and also by an elongation of the phosphorus-nitrogen distance of the polar head group from two to six carbons. Table 2 gives the half lethal dosages (LD_{50}) for the tested compounds after 48 h of incubation. Compared to the ID_{50} values, the LD_{50} values were higher because they represent cell death as a late event in the cytotoxic course of action. Here, in comparison to the other cell lines, HL 60 was most sensitive. In this cell line the LD_{50} and ID_{50} values are almost coincidental showing that growth inhibition may be caused by cell death. The ranking order of the compounds in the dye exclusion assay was comparable to the order obtained in the thymidine assay. Since alkylphosphocholines contain the minimal structural constituents of cytotoxic (alkyl)-lysophosphocholines (ALP), we compared the antiproliferative potency of OcPC with those of two well-known ALP. Table 3 gives a comparison of the (^3H)thymidine incorporation of the most active alkylphosphocholine and ALP represented by OcPC, Oc-Me-G-3-PC and

Table 1. Differential antiproliferative sensitivity of 4 leukemia cell lines to alkylphosphocholines and analogues

Compound	ID_{50} values, μM			
	HL 60	U 937	Raji	K562
HePC	12.3 ± 1.7	7.4 ± 1.2	24.6 ± 3.0	38.1 ± 3.7
OcPC	3.9 ± 0.5	1.1 ± 0.2	7.1 ± 0.9	15.5 ± 1.6
$OcPC_6$	33.0 ± 4.2	4.8 ± 0.6	33.0 ± 3.0	49.8 ± 3.6
OlPC	25.2 ± 2.3	7.3 ± 0.5	30.5 ± 3.4	52.7 ± 7.8
$OlPC_6$	70.0 ± 8.0	30.6 ± 3.2	55.0 ± 4.4	> 100

ID_{50} values were obtained from (^{3}H)thymidine incorporation after 48 h. Data represent the mean ±SD of 4–6 independent experiments done in triplicate.

Table 2. Differential cytotoxic sensitivity of 4 leukemia cell lines to alkylphosphocholines and analogues

Compound	LD_{50} values, μM			
	HL 60	U 937	Raji	K562
HePC	18.9 ± 2.7	39.4 ± 3.7	64.0 ± 7.4	> 100
OcPC	4.6 ± 0.7	11.4 ± 0.9	38.8 ± 4.6	> 100
$OcPC_6$	34.6 ± 4.4	70.6 ± 8.4	> 100	> 100
OlPC	16.0 ± 2.1	68.7 ± 8.2	91.6 ± 5.0	> 100
$OlPC_6$	> 100	> 100	> 100	> 100

LD_{50} values obtained from trypan blue dye exclusion test after 48 h. Data represent the mean ±SD of 4–6 independent experiments done in triplicate.

Table 3. Antiproliferative activity of OcPC, Oc-Me-G-3-PC and BM 41.440 in 4 different leukemia cell lines

Compound	ID_{50} values, μM			
	HL 60	U 937	Raji	K562
OcPC	3.9 ± 0.5	1.1 ± 0.2	7.1 ± 0.9	15.5 ± 1.6
Oc-Me-G-3-PC	1.5 ± 0.4	1.8 ± 0.3	2.8 ± 0.3	8.1 ± 1.0
BM 41.440	2.8 ± 0.5	2.5 ± 0.5	3.0 ± 0.6	14.1 ± 1.1

ID_{50} values were from (^{3}H)thymidine incorporation after 48 h. Data represent the mean ±SD of 3 independent experiments done in triplicate.

Table 4. Hemolysis of human erythrocytes by alkylphosphocholines and analogues and the ALP Oc-Me-G-3-PC

Compound	Concentration μM
HePC	6.0 ± 0.3
OcPC	3.6 ± 0.2
$OcPC_6$	3.9 ± 0.3
Oc-Me-G-3-PC	3.2 ± 0.2
OlPC	12.0 ± 0.4
$OlPC_6$	10.0 ± 0.3

Data represent drug concentrations which caused 50% of hemolysis after 5 min in serum-free medium. Values are mean $\pm$SD of 3 independent experiments.

the thioether-phospholipid BM 41.440. The range of the ID_{50} values for the different cell lines was very narrow, indicating similar activity.

In a second set of experiments, human erythrocytes were exposed to increasing concentrations of alkylphosphocholines and ALP. The 50% hemolysis rates after 5 min of incubation are shown in table 4. To avoid binding of the lipids to serum proteins, erythrocytes were incubated in a serum-free medium. From the data in table 4, it is evident that the hemolytic activity of all tested compounds is also within a narrow range. It is noteworthy that the hemolytic activity was not influenced by modification in the polar part of the molecules. On the other hand, both a reduction in the chain length and an introduction of a double bond in the apolar part of the molecules led to a decrease in hemolytic concentrations of alkylphosphocholines.

Discussion

In recent years increasing interest has been focussed on the antineoplastic activity of ALP in experimental cancer therapy. One of the most important aspects in this area of research was the finding that, in contrast to most cytostatic drugs, ALP acts on the plasma membrane and not on the cell nucleus [1, 10]. Based on investigations on the structure-activity relationships of ALP, a long alkyl chain and a phosphocholine moiety represent the minimal structural constituents responsible for the cytotoxic effects of these compounds. This view is strongly supported by a comparison of OcPC and Oc-Me-G-3-PC. As shown in table 3, both compounds have very similar ID_{50} values with regard to all investigated leukemia cell lines. Moreover, the different alkylphosphocholines and analogues revealed clear dose-response relationships in cells incubated up to 72 h. Starting with OcPC, a shortening of the alkyl chain length (HePC) and an introduction of a double bond (OlPC) was accompanied by a marked reduction in antitumoral activity (tables 1, 2). These structural modifications simultaneously led to a decrease in hemolytic activity (table 4). Based on these results, one might assume that the antitumoral action of alkylphosphocholines is caused at least in part by surface-active properties of the molecules. However, cocultivated erythrocytes remained vital in the cell culture up to the highest concentrations used (100 μM) for 72 h and did not show hemolysis (data not shown). Moreover, an elongation of the phosphorus-nitrogen distance from two to six carbons did not result in a change of the hemolytic activity. On the other hand, as shown in tables 1 and 2, this structural modification increased the ID_{50} and LD_{50} values by factors between 3 and 9. We have also performed hemolysis experiments in the presence of serum. In this case, lytic concentrations are much higher. Thus, the concentrations of alkylphosphocholines which are antineoplastically active are by a factor of about 100 lower than the concentrations necessary for hemolysis. Based on these observations, it is concluded that the antineoplastic activity and the hemolytic activity of ether lipids are not related.

Recently, marked antineoplastic effects of various alkylphosphocholines have been observed in chemically induced autochthonous rat mammary carcinoma [3, 4, 22]. Optimal nontoxic dosages of OcPC, HePC and OlPC were administered orally over a period of 5 weeks. All compounds showed distinct antitumoral activity as demonstrated by high percentages of tumor regression during treatment. However, no antitumoral activity was found in the case of $OlPC_6$ [22]. This is in line with the observation that the compound is only marginally, if at all, active in cell experiments (tables 2, 3).

In conclusion, alkylphosphocholines represent a novel group of compounds with remarkable antineoplastic activity in vitro and in vivo. Certainly, this research is in a very early stage of development and further studies are necessary to clarify the potential significance of these compounds in cancer therapy.

Acknowledgements

The skillful technical assistance of A. Hunger is gratefully acknowledged. The present work was generously supported by the Federal Ministry of Research and Technology (BMFT, Bonn-Bad Godesberg, FRG).

References

1 Eibl H, Unger C: Hexadecylphosphocholine, a new and selective antitumor drug. Cancer Treat Rev 1990;17:233–242.

2 Berdel WE, von Hoff DD, Unger C, Schick HD, Fink U, Reichert A, Eibl H, Rastetter J: Ether lipid derivatives: Antineoplastic activities in vitro and structure-activity relationship. Lipids 1986;21:301–304.

3 Hilgard P, Stekar J, Voegeli R, Engel J, Schumacher W, Eibl H, Unger C, Berger MR: Characterisation of antitumor activity of hexadecylphosphocholine (D18506). Eur J Cancer Clin Oncol 1988;24:1457–1461.

4 Scherf HR, Schuler B, Berger MR, Schmähl D. Therapeutic activity of ET-18-OCH_3 and hexadecylphosphocholine against mammary tumors in BD IV rats. Lipids 1987;22:927–929.

5 Unger C, Damenz W, Fleer EAM, Kim J, Breiser A, Hilgard P, Engel J, Nagel GA, Eibl H: Hexadecylphosphocholine, a new etherlipid analogue. Studies on the antineoplastic activity in vitro and in vivo. Acta Oncol 1989;28:213–217.

6 Unger C, Eibl H: Phospholipide als Antitumormittel, zytotoxische Prinzipien als Ergebnis von Strukturvariationen; in Nagel GA (ed): Mammacarcinome – neue Perspektiven experimenteller und Klinischer Therapieforschung. Berlin, Springer, 1986, pp 115–123.

7 Honma Y, Kasukabe T, Hozumi M, Tsushima S, Nomura H: Induction of differentiation of cultured human and mouse myeloid leukemia cells by alkyllysophospholipids. Cancer Res 1981;41:3211–3216.

8 Munder PG, Weltzien HU, Modolell M: Lysolecithin analogs: A new class of immunopotentiators; in Miescher PA (ed): VII Int Symp Immunopathology. Basel, Schwabe, 1976, pp 411–424.

9 Storme G, Berdel WE, Bruyneel R, De Bruyne G, Mareel M: Antiinvasive effects of an alkyl-lysophospholipid ET-18-OCH_3 which has no microtubule inhibitor activity. Proc Am Assoc Cancer Res 1983;24:31.

10 Baumann WJ (ed): First International Symposium on Ether Lipids in Oncology. Lipids 1987;22:787–974.

11 Unger C, Eibl H, von Heyden HW, Nagel GA: A new assay for the O-alkyl cleavage enzyme with alkyl-lysophospholipids as substrates. Cancer Res 1985;45:616–618.

12 Kötting J, Unger C, Eibl H: Substrate specificity of O-alkylglycerol monooxygenase (E.C. 1.14.16.5), solubilized from rat liver microsomes. Lipids 1987;22:831–835.

13 Neumüller W, Fleer EAM, Unger C, Eibl H. Enzymatic acylation of ether and ester lysophospholipids in rat liver microsomes. Lipids 1987;22:808–812.

14 Unger C, Eibl H, Kim DJ, Fleer EAM, Bartsch HH, Nagel GA, Pfitzenmaier K: Sensitivity of leukemic cell lines to cytotoxic alkyllysophospholipids in relation to alkyl-cleavage enzyme activities. J Natl Cancer Inst 1987;78:219–222.

15 Eibl H, Unger C: Phospholipide als Antitumormittel: Möglichkeiten einer selektiven Therapie; in Unger C, Eibl H, Nagel GA (eds): Die Zellmembran als Angriffspunkt der Tumortherapie. Aktuel Onkol 1987;34:1–16.

16 Unger C: Alkylphosphocholine und Analoga: Entwicklung einer Substanzgruppe mit antineoplastischer Wirkung; Habschr, Universität Göttingen, 1987.

17 Dimbeck W: Künstliche Phospholipide mit vergrössertem Phosphorstickstoff-Abstand; Diss, Universität Braunschweig, 1976.

18 Eibl H, Woolley P: A general synthetic method for enantiomerically pure ester and ether lysophospholipids. Chem Phys Lipids 1988;47:63–68.

19 Eibl H: Phospholipide als funktionelle Bausteine biologischer Membranen. Angew Chemie 1984;23:257–268.
20 Eibl H: Synthesis of phospholipids; in Knight CG (ed): Liposomes, from Structure to Therapeutic Application. Amsterdam, Elsevier, 1981, pp 19–38.
21 Reman F, Demel R, De Gier J, Van Deenen LLM, Eibl H, Westphal O: Studies on the lysis of red cells and bimolecular lipid leaflets by synthetic lysolecithins, lecithin structural analogs. Chem Phys Lipids 1969;3:221–233.
22 Berger MR, Schmähl D, Unger C, Eibl H: Antineoplastic activity of alkylphosphocholine derivatives in methylnitrosourea (MNU)-induced rat mammary carcinoma. Proc Am Assoc Cancer Res 1988;27:330.

Clemens Unger, Medizinische Universitätsklinik, Abteilung Hämatologie/Onkologie, D-W-3400 Göttingen (FRG)

Eibl H, Hilgard P, Unger C (eds): Alkylphosphocholines: New Drugs in Cancer Therapy. Prog Exp Tumor Res. Basel, Karger, 1992, vol 34, pp 33–46

Cellular Uptake and Metabolic Fate of Hexadecylphosphocholine

E.A.M. Fleer[a], *D. Berkovic*[a], *C. Unger*[a], *H. Eibl*[b]

[a]Medizinische Universitätsklinik, Abteilung Hämatologie/Onkologie, and
[b]Max-Planck-Institut für biophysikalische Chemie, Göttingen, FRG

Introduction

Since about three decades, the antineoplastic and immune modulating action of lysophosphatidylcholines (LPCs) and ether-linked lysophosphatidylcholines (ALPs) has been described [1–6] and structure-activity relationship studies have shown that the lysophospholipid character of these compounds (fig. 1) is of crucial importance for their antineoplastic action [5]. For instance an ester-linked LPC can be readily metabolized and is not an active anticancer agent although early reports showed its immune-modulating properties [1–3]. Also, the sn-2 position of the ALP must be blocked to prevent detoxification by (a) acylation or (b) degradation by 1-0-alkyl-glycerol monooxidase (AGMO), an enzyme able to split the molecule by oxidation of the ether bond [7–10]. In studies on the minimal requirements of ALPs and related compounds for their antitumoral action, alkylphosphocholines (APCs) emerged as a new class of antineoplastic drugs [17–21], a summary of the results of these studies has been published recently [11]. This group of compounds, of which hexadecylphosphocholine (HePC) is the most prominent representative, shows anticancer properties similar as the well-known ALPs ET18OCH3 (1-0-octadecyl-2-0-methyl-rac-glycero-3-phosphocholine) [12–15] and BM 41.440 (1-S-hexadecyl-2-0-methoxy-methyl-propananol-3-phosphocholine) [16].

HePC, for instance, shows strong cytotoxic action in vitro [17] and, in treatment of chemically induced tumors in mice and rats, selective antineoplastic activity in vivo [18–21]. In cell culture experiments, an important property of LPCs and analogues is their tendency to form micelles [22–25]. If no other mechanism is involved, cells will be damaged at concentrations

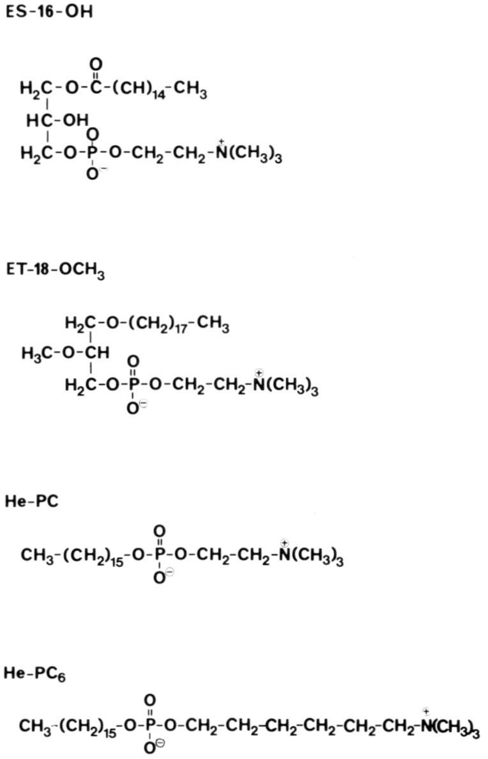

Fig. 1. The chemical structure of ester LPCs, ALPs, and APCs.

above the critical micellar concentration (CMC) [26]. However, LPCs [27–33] and HePC [34] are bound to serum proteins. This is in agreement with earlier literature, where it was reported that HePC may bind to lipid-metabolizing enzymes such as phospholipase A_2 [24, 35, 36] in a similar way as LPCs [24].

ALPs and APCs show cytotoxic effects at concentrations far below the point where they induce hemolysis [12, 37], strongly suggesting that mechanisms other than mere physicochemical effects are involved in the antineoplastic activity of APCs. Furthermore, cytotoxic effects of HePC are found at largely varying concentrations for cells from different cell lines [Berkovic et al., this volume], showing that these cells have large differences in sensitivity for HePC.

In this chapter we will describe how HePC is incorporated in the cells of cultured human cell lines. Subsequently, data on the metabolism of HePC in human lymphomic Raji cells are discussed and compared with data of specific HePC analogues that can only be metabolized in a defined site of the molecule.

Cellular Uptake of HePC

One reason for the varying sensitivity of different cell lines for HePC may be a difference in the incorporation of this compound in cellular membranes, e.g. sensitive cells take up HePC at a faster rate or at higher amounts (or both) than less-sensitive cells. A problem in defining the uptake of HePC in different cells is the fact that cells from different tumors or transformed cell lines may vary in size over a wide range. In order to circumvent this problem, and to get at least a crude measure of normalisation, we decided to compare the uptake of HePC per 100 nmol of cellular phospholipids. In table 1, the phospholipid contents of the different cells tested are listed and a correlation between phospholipid content and cell size seems to exist. For instance, HL60 cells, which are the most sensitive and smallest of the lymphatic cells tested, have a similar phospholipid content as, for instance, the slightly larger lymphomic Raji cells, but the rather large erythroblastic K562 cells show higher phospholipid contents. Also, the size difference between KB, MDA MB 231 and 3T3 cells, all adherent cells, is not very large, corresponding well to an only slight difference in phospholipid content. In figure 2 the uptake of HePC per 100 nmol of cellular phospholipid is plotted as a function of time at 37 °C. From the data it is evident that all cells take up HePC. With the exception of KB cells, which are the most sensitive of all cells tested, incorporation of HePC seems to occur at similar rates and, within a reasonable margin, to the same extent when intake is measured over a 5-hour time period. For instance, the incorporated amount of HePC in the very sensitive HL60 cells and the amount taken up by the almost completely insensitive K562 or 3T3 cells is very similar with an uptake of 4–5 nmol per 100 nmol of cellular phospholipid. However, Raji cells, which are far less sensitive than HL60 cells, and U937 cells incorporate more HePC in their cellular membranes. This seems to illustrate that the amount of HePC in the different cells – and its concentration in the cellular membranes – is not a measure for the sensitivity of a specific cell line towards HePC. Probably a more specific interaction than the accumulation in phospholipid membranes is involved in the action of HePC on some cell lines and its relative lack of activity on other cell lines.

Table 1. Cellular lipid content[1]

Cell line	Lipid phosphorous nmol/10^6 cells	Cholesterol nmol/10^6 cells	Ratio[2]
KB	108.8 ± 13.1	16.0 ± 4.8	6.8
MDA MB 231	110.8 ± 11.5	24.4 ± 7.5	4.5
HL 60	40.7 ± 7.1	7.4 ± 3.0	5.5
K 562	65.1 ± 5.2	17.9 ± 2.4	3.6
U 937	38.0 ± 5.5	10.1 ± 1.5	3.7
Raji	35.5 ± 2.0	8.7 ± 1.8	4.1
Swiss 3T3	143.9 ± 14.4	29.5 ± 18.3	4.9

[1] Lipid phosphorous [61] and cholesterol content [62] were determined from lipid extracts [63] of a known number of cells. Data represent the mean ± SD of at least five independent determinations.
[2] The ratio of lipid phosphorous over cholesterol, calculated from the mean values.

At lower temperatures, the uptake of HePC in the cells is increasingly inhibited. Whereas at 37 °C HePC is rapidly incorporated in the cells, at 20 °C uptake is reduced to less than 50% and after incubations at 4 °C only traces of HePC (less than 2%) are found in the cells [36]. This seems to indicate that at 4 °C HePC is no longer internalized and the small amounts of HePC that are still connected to the cells might be due to binding to extracellular membrane proteins or to insertion into the outer leaflet of the plasma membrane phospholipid bilayer.

To distinguish between these possibilities, concentration-dependency studies of HePC binding to cells at 4 °C were performed and no sign of saturation of binding was observed in a HePC concentration range from 1 to 100 μg/ml (ca. 2.5–250 μ*M*), indicating that HePC binding to the cell outer membrane is nonspecific and no receptor protein for HePC is accessible in the cellular plasma membrane. As reported before HePC binds to serum proteins [34]. This binding to serum proteins should decrease the amount of HePC available for cellular uptake by lowering the concentration of free HePC in the medium. By comparing the uptake of HePC in the presence and absence of FCS, it is clear that 10% FCS indeed lowers HePC uptake rates in the four cell lines tested, to about 50% of its value in absence of FCS. That this effect is not more pronounced seems to correlate with the dynamic state of this system. Binding of HePC to serum proteins is not an absolute, high-affinity process. When the equilibrium between free HePC and serum-bound HePC is disturbed,

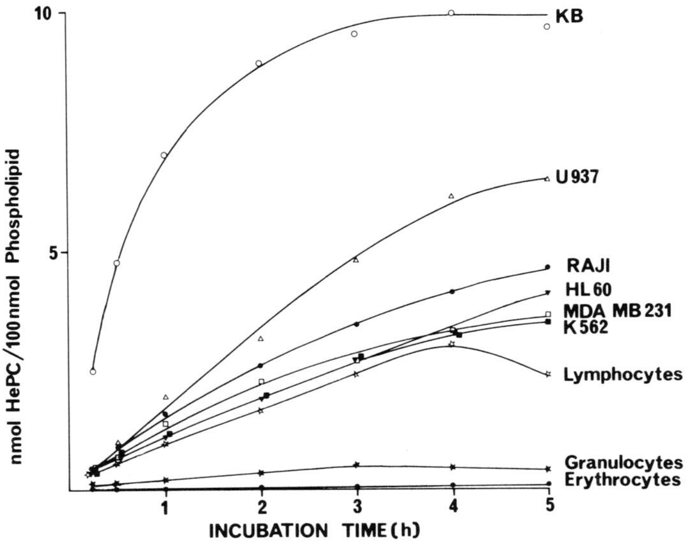

Fig. 2. Uptake of HePC in different cells and cell lines at 37 °C. Uptake was determined from the amount of radiolabel incorporated in the cells after washing three times with medium containing 10% fetal calf serum to remove aspecifically bound HePC. The uptake was calculated per 100 nmol cellular phospholipids using the known number of cells and the phospholipid content of the cells given in table 1. The cells were incubated for 2 h at 37 °C after trypsination (adherent cells) and dilution to the desired cell count. The uptake experiment was started by the addition of HePC, radiolabeled in the apolar chain. At indicated time points, aliquots of the cells were taken to measure the amount of radiolabel that was incorporated. After 2, 3, 4 and 5 h of incubation, aliquots were taken for cell counts to establish that cells were at least 90% viable throughout the incubation period. Incubation conditions: RPMI 1640 medium supplemented with 10% fetal calf serum, 2 m*M* glutamine, 100 IU penicillin and 100 μg streptomycin per ml, 37 °C.

for instance by reducing the amount of free HePC through binding to the cell's plasma membrane, HePC is immediately released from serum proteins to establish a new equilibrium. For small changes in the free HePC concentration, this system is similar to a buffer system, where FCS is used to keep HePC at a low but more or less constant concentration (fig. 3). It is not clear whether HePC must be free in solution to be inserted in the plasma membrane of the cells or if it is possible that a direct transfer from serum proteins to the plasma membrane also takes place.

When cells are incubated with energy depletors like sodium azide,

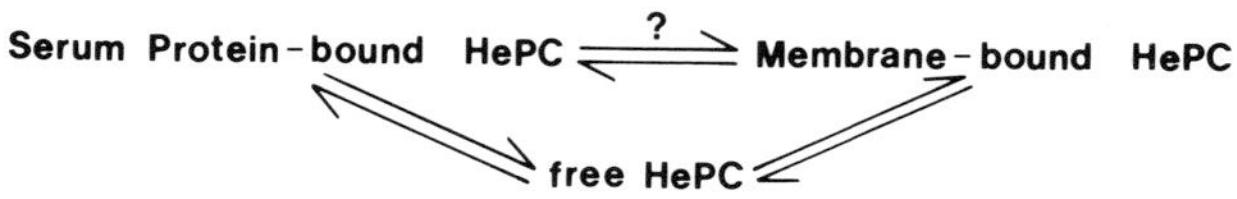

Fig. 3. Equilibrium diagram for HePC free in solution, bound to serum proteins and bound to plasma membranes. Since it is not clear if serum proteins can help to insert HePC directly into the plasma membranes of cells a questionmark is added to state this uncertainty.

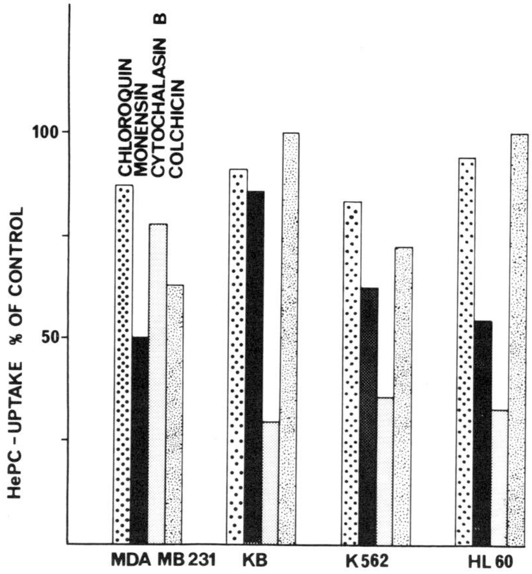

Fig. 4. Uptake of HePC in the presence of several endocytotic inhibitors in two HePC-sensitive (HL60, lymphatic, and KB, adherent) cell lines and two HePC-resistant (K562, lymphatic and MDA MB231, adherent) cell lines. Cells were preincubated at 37 °C for 2 h with the respective drug and HePC uptake was started by the addition of radiolabeled HePC as described in figure 2. The cells were harvested 2 h after addition of radiolabeled HePC. Data are presented as percentage of uptake in the absence of endocytotic inhibitors in the same cell line. Concentrations: monensin 20 μg/ml (29.8 *μM*), chloroquin 20 μg/ml (62.5 *μM*), cytochalasin B 20 μg/ml (41.8 *μM*) and colchicin 1.5 μg/ml (3.75 *μM*). RPMI 1640 medium was complemented with 10% FCS, glutamine, penicillin and streptomycin as given in figure 2. At the end of the experiment additional aliquots were taken to establish that the cells were at least 90% viable throughout the incubation period.

desoxyglucose or sodium thiocyanate, HePC uptake is hardly affected [36]. However, when cells are incubated with cytotoxic compounds like chloroquin, monensin, cytochalasin B or colchicin, inhibition of HePC incorporation can be observed (fig. 4). Different cells show different behavior, for instance in MDA MB 231, cytochalasin B is only moderately effective with a 20% decrease of uptake, while colchicine (30%) and monensin (50%) are very effective. In contrast, in the KB cell line, cytochalasin B is the most potent inhibitor of HePC uptake, with a reduction of 70%. Monensin is far less effective and only slightly more active than choloroquine, with 15% inhibition of uptake.

Proposal for Possible HePC Uptake Mechanisms

Uptake of HePC is dependent on concentration, time and temperature. At a concentration of 10 μg/ml (24.5 μ*M*), half-time of saturation at 37 °C is between 30 and 60 min, and it seems that uptake is only slightly dependent on metabolic energy. From the fact that HePC incorporation is temperature dependent and is blocked partly by compounds known to have an inhibiting effect on endocytotic processes [38–44], HePC may be taken up by an endocytotic process, such as pinocytosis, phagocytosis or membrane renewal. It is possible, and, given the physicochemical properties of HePC, even likely, that part of the HePC that is incorporated in the cell is first inserted into the outer leaflet of the plasma membrane before being transported into the intracellular compartments. On the other hand, nonspecific binding of HePC on cell surface proteins followed by internalization, or even flip-flop [45–47] of HePC to the inner leaflet of the plasma membrane cannot be excluded on the basis of the above-mentioned data. Still, if flip-flop would play a crucial role in the transport of HePC over the plasma membrane of the cell, HePC uptake should show a concentration-dependent saturation behavior and probably some type of specific binding saturation at a nonpermissive temperature, which were not observed in our experiments. Therefore, the most likely mechanism for HePC uptake into the intracellular compartments seems, in our view, to be an insertion in the outer leaflet of the plasma membrane followed, for instance, by membrane renewal.

Metabolism of HePC

Metabolism of HePC was studied using the human lymphomic Raji cell line (B lymphoblastic lymphoma). For metabolic studies, HePC labeled in the choline-methyl groups was used. It was found that HePC is degraded very slowly and that a major pathway for degradation seems to be a transfer of the (phospho)choline group to diacylglycerol, yielding (diacyl)phosphatidylcholine and, probably, hexadecanol [17]. This transfer reaction was found earlier for OcMeG3PC, labelled in the polar headgroup [48] re-affirming data of other groups with OcMeG3PC labelled in the octadecyl chain of the molecule, where 1-0-octadecyl-2-0-methyl glyerol was found, as is expected from the above transfer reaction [7, 49]. Because no radiolabel was detected in the water phases of Bligh and Dyer cell extracts, it was suggested that the transfer of the phosphocholine head group is a direct mechanism. This mechanism seems to be similar to the mechanism described by Voelker and Kennedy [50, 51], where the transfer of the phosphocholine group from phosphatidyl choline to ceramide was described. This reaction yields diacylglycerol as one product and sphingomyelin as the other. These authors also showed a direct transfer of the phosphocholine head group, and not a hydrolysis of the phosphate ester bond followed by the utilization of phosphocholine in a (diacyl)phosphatidylcholine formation reaction.

Furthermore, it was observed that the amount of phosphatidylcholine formed in the transfer reaction seemed to correlate well with the percentage of cell death in trypan blue dye exclusion assays [17]. This observation led to the hypothesis that HePC might be a prodrug that is potentiated by the above-mentioned reaction. On the other hand, it is also possible that the intracellular location where this reaction is taking place is a 'target organelle' for HePC. To distinguish between these possibilities, we have later performed incubation experiments on Raji cells with phosphono analogues of HePC, where either the possibility for degradation by phospholipase C- or by phospholipase D-like enzymes was inhibited (Ries et al., in preparation] (fig. 5). These phosphono analogues were not degraded at all by Raji cells in metabolic experiments lasting up to 3 days. However, both types of analogue showed cytostatic and cytotoxic properties in Raji cells as well as in U937 cells, with a slightly lower efficiency. In addition, all the phosphono analogues showed a certain degree of antitumoral activity in rats bearing chemically induced mammary carcinomas, although none of them was as efficient as HePC. From these observations we now conclude that degradation of HePC is not a prerequisite for its action on tumor cells.

Concluding Remarks

In this chapter we have described some aspects of hexadecylphosphocholine activity on a number of selected cell lines, and discussed how these effects are related to the physicochemical properties of HePC. It seems clear, from the data presented, that although HePC is a potential cytotoxic agent to all cells tested, there are large differences in the sensitivity of these cells for HePC. These differences in HePC sensitivity are not adequately

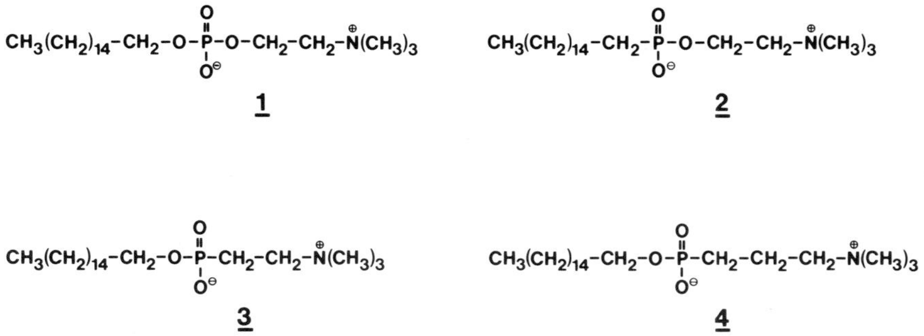

Fig. 5. Structure of phosphonoanalogs of HePC. 1 = Hexadecylphosphocholine, 2 = cholinyl hexadecylphosphonate (inaccessible for phospholipase C); 3 = hexadecyl cholinylphosphonate (inaccessible for phospholipase D); 4 = hexadecyl(N,N,N,-trimethylammonio)propyl phosphonate (inaccessible for phospholipase D).

explained by either the physicochemical properties of HePC or by differences in uptake of HePC in the cells. Also, it seems that degradation of HePC by phospholipase C- or phospholipase D-like enzymes is not giving a sufficient explanation for the observed differences in HePC sensitivity. Therefore, more specific, cell-based differences are thought to be responsible for the effects of HePC on the cells. One possible effect of HePC, responsible at least for inhibition of cell growth, might be the induction of cell differentiation as observed in U937 cells [52] and in HL60 cells [21], or the effect of ALPs on internal signal transduction pathways, for instance the inhibition or stimulation of protein kinase C [34]. However, at least in the latter case, it is not clear if the influence of protein kinase C in cell survival is not overestimated, since conflicting data on the role of protein kinase C have been reported [53–59].

Also, how the signal transduction is transported from the plasma membrane to the cell nucleus is not yet clearly understood, as is the role of the internal calcium concentration or the frequently observed transient rise in the internal pH, when cells are stimulated. It is thus that we can only have an incomplete understanding of the role that HePC might play in any of these possible pathways. Therefore, we think that in order to be able to positively identify which of the effects of HePC in the cells is responsible for its observed cytotoxic action (as well as its in vivo antineoplastic activity), we probably need to know a great deal more, not only about where HePC is influencing important cellular control mechanisms, but also how these mechanisms are controling vital events in the cell.

Additionally, it has to be kept in mind that the data that are generated in cell culture experiments may give a very detailed view on the activity of HePC on a cellular level, but that its antineoplastic activity can only be evaluated from experiments with animal models. That is, the effects of HePC on a complete organism, such as a mouse or rat, may differ to a great extent from the effects of HePC on a single cell line.

For HePC the data obtained from cell culture experiments could be reaffirmed in experimental treatment of chemically induced mammary carcinomas in the rat [18–20], perhaps because of the fact that HePC is only slowly degraded [60] and shows rather long half-life times in rats and mice [37].

Finally, caution must be taken to extrapolate the experimental data to the treatment of cancer in human beings. It cannot be excluded that, although HePC was successful in the experimental models, its activity in human beings might be limited. However, from preliminary data, it seems that at least in topical application schedules on skin metastases of mammary carcinomas and lymphomas, HePC might be a successful anticancer drug [Unger et al., and Dummer et al., this vol.].

Surely, HePC, as one of the first lipid-derived antineoplastic agents brought from the experimental stage to clinical trial studies, may turn out not to be the ultimate antitumor drug and the search for better, more efficient agents with even less side effects should continue. In this search, we think that a good understanding of the cellular events that are influenced by HePC may be of great help in identifying new and more powerful alkylphosphocholine analogues.

References

1 Burdzy K, Munder PG, Fisher H, Westphal O: Steigerung der Phagozytose von Peritonealmakrophagen durch Lysolecithin. Z Naturforsch 1964;19b:118–120.

2 Elsbach P, van den Berg JWO, van den Bosch H, van Deenen LIM: Metabolism of phospholipids by polymorphonuclear leukocytes. Biochem Biophys Acta 1965;106:338–347.

3 Munder PG, Modollel M: Adjuvant induced formation of lysophosphatides and their role in immune response. Int Arch Allergy Apply Immunol 1973;45:133–135.

4 Munder PG, Weltzien HU, Modollel M: Lysolecithin analogs: A new class of immunopotentiators; in Miescher PA (ed): Int Symp Immunopathology. Basel, Schwabe Publishers, 1976, pp 411–424.

5 Munder PG, Berchtold R, Pahlke W: Tumorzellzerstoerung durch Alkylphospholipide: Beziehung zwischen chemischer Struktur und biologischer Aktivitaet; in Unger C, Eibl H, Nagel GA (eds): Die Zellmembran als Angriffspunkt der Tumortherapie. München, Zuckschwert Verlag, 1987, pp 37–66.

6 Yamamoto N, Ngwenya BZ, Sery TW, Pieringer RA: Activation of macrophages by ether analogues of phospholipids. Cancer Immunol Immunother 1987;25:185–192.

7 Snyder F, Record M, Smith Z, Blank ML, Hoffman DR: Selective cytostatic action of ether-lipid analogs of PAF: Mechanistic studies related to their metabolism, subcellular localization, and effects on cellular transport systems; in Unger C, Eibl H, Nagel GA (eds): Die Zellmembran als Angriffspunkt der Tumortherapie. München, Zuckschwert Verlag, 1987, pp 19–26.

8 Neumueller W, Fleer EAM, Unger C, Eibl H: Enzymic acylation of ether and ester lysophospholipids in rat liver microsomes. Lipids 1987;22:808–812.

9 Soodsma JF, Piantadosi C, Snyder F: The biocleavage of alkyl glyceryl ethers in Morris hepatomas and other transplantable neoplasms. Cancer Res 1970;30:309–311.

10 Koetting J, Unger C, Eibl H: Substrate specificity of O-alklyglycerolmonooxygenase (EC 1.14.16.5), solubilized from rat liver microsomes. Lipids 1987;22:831–835.

11 Fleer EAM, Kim D-J, Eibl H, Unger C: Cytotoxic activity of lysophosphatidylcholine analogues on human lymphoma Raji cells. Onkologie 1990;13:295–300.

12 Weltzien HU, Westphal O: Synthesen von Cholinphosphatiden IV: O-methylierte und O-acetylierte Lysolecithine. Liebigs Ann Chem 1967;709:240–243.

13 Storch J, Munder PG: Increased membrane permeability for an antitumoral alkylphospholipid in sensitive tumor cells. Lipids 1987;22:813–819.

14 van Blitterswijk WJ, Hilkman H, Storme GA: Accumulation of an alkyllysophospholipid in tumor cell membranes affects membrane fluidity and tumor cell invasion. Lipids 1987;22:820–823.

15 Danhauser S, Berdel WE, Schick HD, Fromm M, Reichert A, Fink U, Busch R, Eibl H, Rastetter J: Structure-cytotoxicity studies on alkyl lysophospholipids and some analogs in leukemic blasts of human origin in vitro. Lipids 1987;22:911–915.

16 Fromm M, Berdel WE, Schick HD, Fink U, Pahlke W, Bicker U, Reichert A, Rastetter J: Antineoplastic activity of the thioether lysophospholipid derivative BM 41.440 in vitro. Lipids 1987;22:916–918.

17 Fleer EAM, Unger C, Kim D.-J, Eibl H: Metabolism of ether phospholipids and analogs in neoplastic cells. Lipids 1987;22:856–861.

18 Muschiol C, Berger MR, Schuler B, Scherf HR, Garzon FT, Zeller WJ, Unger C, Eibl H, Schmähl D: Alkylphosphocholines: Toxicity and anticancer properties. Lipids 1987;22:930–934.

19 Berger MR, Schmähl D, Eibl H, Unger C: Structure-activity relationship of alykylphosphocholines in methylnitrosourea (MNU)-induced rat mammary carcinoma; in Nagel GA, Kornhuber B (eds): Cancer Research and Clinical Oncology (suppl to vol 114). Proc 19th National Cancer Congr German Cancer Society. Berlin, Springer, 1988, vol 114 p 73.

20 Hilgard P, Stekar J, Voegeli R, Engel J, Schumacher W, Eibl H, Unger C, Berger MR: Characterization of the antineotumor activity of hexadecylphosphocholine (D 18506). Eur J Cancer Clin Oncol 1988;24:1457–1461.

21 Hilgard P, Harlemann JH, Voegeli R, Maurer HR, Echanti D, Unger C: The antineoplastic activity of hexadecylphosphocholine (HPC) is associated with tumor cell differentiation. Proc Am Assoc Cancer Res 1989;30:2310.

22 Haydon DA, Taylor J: The stability and properties of bimolecular lipid leaflets in aqueous solutions. J Theor Biol 1963;4:281–296.

23 Reman F, Demel R, De Gier J, Van Deenen LLM, Eibl H, Westphal O: Studies on the lysis of red cells and biomolecular lipid leaflets by synthetic lysolecithin, lecithin structural analogs. Chem Phys Lipids 1969;3:221–233.

24 van Dam-Mieras MCE, Slotboom AJ, de Haas GH: The interaction of phospholipase A_2 with micellar interfaces. The role of the N-terminal region. Biochemistry 1975;14:5387–5394.

25 Pascher I, Sundell S, Eibl H, Harlos K: The single crystal structure of 1-octadecyl-2-methyl-glycero-phosphocholine monohydrate. A multilamellar structure with interdigitating headgroups and hydrocarbon chains. Chem Phys Lipids 1986;36:53–64.

26 Ray TK, Skipski VP, Barclay M, Essner E, Archibald FM: Lipid composition of rat liver plasma membranes. J Biol Chem 1969;224:5528–5536.

27 Klopfenstein WE: Enthalpy change of binding of lysolecithin to serum albumin. Biochem Biophys Acta 1969;181:323–325.

28 Baisted DJ, Robinson BS, Vance DE: Albumin stimulates the release of lysophosphatidylcholine from cultured rat hepatocytes. Biochem J 1988;253:693–701.

29 Sekas G, Patton GM, Lincoln EC, Robins SJ: Origin of plasma lysophosphatidylcholine: Evidence for direct hepatic secretion in the rat. J Lab Clin Med 1985;105:190–194.

30 Robinson BS, Baisted DJ, Vance DE: Comparison of albumin-mediated release of lysophosphatidylcholine and lysophosphatidylethanolamine from cultured rat hepatocytes. Biochem J 1989;264:125–131.

31 Graham A, Bennett AJ, McLean AAM, Zammit VA, Brindley DN: Factors regulating the secretion of lysophosphatidylcholine by rat hepatocytes compared with the synthesis and secretion of phosphatidylcholine and triacylglycerol. Effects of albumin, cycloheximide, verapamil, EGTA, and chlorpromazine. Biochem J 1988;253:687–692.

32 Robinson BS, Yao Z, Baisted DJ, Vance DE: Lysophosphatidylcholine metabolism and lipoprotein secretion by cultured rat hepatocytes deficient in choline. Biochem J 1989;268:207–214.

33 Switzer S, Eder HA: Transport of lysolecithin by albumin in human and rat plasma. J Lipid Res 1965;6:506–511.

34 Ueberall F, Oberhuber H, Maly K, Zaknun J, Demuth L, Grunicke H: Hexadecylphosphocholine inhibits inositolphosphate formation and protein kinase C activity. Cancer Res 1991;51:807–812.

35 Fleer EAM, Berkovic D, Eibl H, Unger C: Cellular uptake of hexadecylphosphocholine (He-PC). Proc Am Assoc Cancer Res 1990;31:414.

36 Berkovic D, Fleer EAM, Hilgard P, Eibl H, Unger C: Serum concentration of hexadecylphosphocholine (He-PC) in the rat. Proc Am Assoc Cancer Res 1990;31:413.

37 Fleer EAM, Puijk WC, Slotboom AJ, de Haas GH: Modification of arginine residues in porcine pancreatic phospholipase A2. Eur J Biochem 1981;116:277–284.

38 Unger J, Ketelbrant P: Inhibition by monensin of dog thyroid secretion in vitro. Endocrinology 1988;123:66–71.

39 Hedin U, Thyberg J: Receptor-mediated endocytosis of immunoglobulin-coated colloidal gold particles in cultured mouse peritoneal macrophages. Chloroquine and monensin inhibit transfer of the ligand from endocytotic vesicles to lysosomes. Eur J Cell Biol 1985;39:130–135.

40 Johnson DC, Spear PG: Monensin inhibits the processing of herpes simplex virus glycoproteins, their transport to the cell surface, and the egress of virions from infected cells. J Virol 1982;43:1102–1112.

41 Poole B, Ohkuma S: Effect of weak bases on the intralysosomal pH in mouse peritoneal macrophages. J Cell Biol 1981;90:665–669.

42 Berg T, Blomhoff R, Naess L, Tolleshaug H, Drevon, CA: Monensin inhibits receptor-mediated endocytosis of asaloglycoproteins in rat hepatocytes. Exp Cell Res 1983;148:319–330.

43 Storrie B, Pool RR Jr, Sachdeva M, Maury KM, Oliver C: Evidence for both prelysosomal and lysosomal intermediates in endocytotic pathways. J Cell Biol 1984;98:108–115.

44 Thirion J, Wattiaux R: Effect of monensin on fluid phase and receptor mediated endocytosis by rat hepatocyte monolayers. Biochem Biophys Res Commun 1988;152:1275–1281.

45 Renooy W, van Golde LMG: The transposition of molecular classes of phosphatidylcholine across the rat erythrocyte membrane and their exchange between the red cell membrane and plasma lipoproteins. Biochem Biophys Acta 1977;470:465–474.

46 Rousselet A, Guthmann C, Matricon J, Bienvenue A, Devaux P: Study of the transverse diffusion of spin labeled phospholipids in biological membranes. I. Human red blood cells. Biochem Biophys Acta 1976;426:357–371.

47 van Meer G, Poorthuis JHM, Wirtz KWA, Op Den Kamp, JAF, Van Deenen LLM: Transbilayer distribution and mobility of phosphatidylcholine in intact erythrocyte membrances. a study with phosphatidylcholine exchange protein. Eur J Biochem 1980;103:283–288.

48 Fleer EAM, Kim D-J, Unger C, Eibl H: 1-O-Octadecyl-2-O-methyl-rac-glycero-3-phospho(^{3}H-methyl]-choline: Chemical preparation and metabolism in leukemic Raji cells; in Muccino RR (ed): Synthesis and Application of Isotopically Labeled Compounds. Proc 2nd Int Symp, Kansas City, September 3–6. Amsterdam, Elsevier, 1985, pp 473–478.

49 van Blitterswijk WJ, van der Bend RL, Kramer IJM, Verhoeven AJ, Hilkman H, De Widt J: A metabolite of an antineoplastic ether phospholipid may inhibit transmembrane signaling via protein kinase C. Lipids 1987;22:842–846.

50 Voelker DR, Kennedy EP: Cellular and enzymic synthesis of sphingomyelin. Biochemistry 1982;21:2753–2759.

51 Voelker DR, Kennedy EP: Phospholipid exchange protein-dependent synthesis of sphingomyelin. Meth Enzymol 1983;98:596–598.

52 Hochhuth C, Berkovic D, Eibl H, Unger C, Doenecke D: Effects of antineoplastic phospholipids on parameters of cell differentiation in U937 cells. J Cancer Res Clin Oncol 1990;116;459–466.

53 Sigmund O, Naor Z, Anderson DJ, Stein R: Effect of nerve growth factor and fibroblast growth factor on SCG10 and c-fos expression and neurite outgrowth in protein kinase C-depleted PC12 cells. J Biol Chem 1990;265:2265–2261.

54 Levin DE, Fields FO, Kinishawa R, Bishop JM, Thorner J: A candidate protein kinase C gene, PKC1, is required for the *S. cerevisiae* cell cycle. Cell 1990;62:213–224.

55 Kraft AS, Wagner F, Housey GM: Overexpression protein kinase C beta 1 is not sufficient to induce factor independence in the interleukin-3-dependent myeloid cell line FDC-P1. Oncogene 1990;5:1243–1246.

56 Choi PM, Tchou-Wong KM Weinstein IB: Overexpression of protein kinase C in HT29 colon cancer cells causes growth inhibition and tumor suppression. Mol Cell Biol 1990;19:4650–4657.

57 Eldar H, Zisman Y, Ullrich A, Livneh E: Overexpression of protein kinase C alpha-subtype in Swiss/3T3 fibroblasts causes loss of both high and low affinity receptor numbers for epidermal growth factor. J Biol Chem 1990;265:13290–13296.

58 Isakov N, McMahon P, Altman A: Selective post-transcriptional down-regulation of protein kinase C isozymes in leukemic T cells chronically treated with phorbol ester. J Biol Chem 1990;265:2091–2097.

59 Cuadro A, Molloy CJ, Pech M: Expression of protein kinase C in NIH 3T3 cells increases its growth response to specific activators. FEBS Lett 2990;260:281–284.

60 Breiser A, Kim D-J, Fleer EAM, Damenz W, Drube A, Berger MR, Nagel GA, Eibl H, Unger C: Distribution and metabolism of hexadecylphosphocholine in mice. Lipids 1987;22:925–926.

61 Eibl H, Lands WE: A new, sensitive determination of phosphate. Anal Biochem 1969;30:51–57.
62 Sale FO, Marchesini S, Fishman PH, Berra BA: Sensitive enzymatic assay for determination of cholesterol in lipid extracts. Anal Biochem 1984;145:347–350.
63 Bligh EG, Dyer WJ: A rapid method of total lipid extraction and purification. Can J Biochem Physiol 1959;37:911–917.

E.A.M. Fleer, Medizinische Universitätsklinik, Abteilung Hämatologie/Onkologie, D-W-3400 Göttingen (FRG)

Eibl H, Hilgard P, Unger C (eds): Alkylphosphocholines: New Drugs in Cancer Therapy. Prog Exp Tumor Res. Basel, Karger, 1992, vol 34, pp 47–58

Morphological Changes of Adherent and Nonadherent Cells by Treatment with Hexadecylphosphocholine and 1-O-Octadecyl-2-O-methyl-*rac*-glycero-3-phosphocholine Observed by Scanning Electron Microscopy

P. Kaufmann-Kolle, E.A.M. Fleer[1]

Max-Planck-Institut für biophysikalische Chemie, Abteilung Membranbiophysik, Göttingen, FRG

Introduction

Alkylphosphocholines and (alkyl)-lysolecithins have remarkable antineoplastic activity. These antitumoral effects include the activation of macrophages [1], inhibition of tumor cell invasion into normal tissues [2], induction of differentiation of malignant cells [3], and also direct cytotoxicity to neoplastic cells in vitro [4–6] and in vivo [7, 8].

It is conceivable, and confirmed by different authors, that the plasma membrane of cells is a major target for these new antineoplastic agents [9, 10]. It is well documented that lipids with lytic properties, such as 1-O-alkyl-glycero-3-phosphocholines, can damage cells only in the absence of serum or plasma proteins in concentrations at or above the critical micellar concentration (CMC). In addition, 1-O-octadecyl-2-O-methyl-*rac*-glycero-3-phosphocholine (OcMeG3PC) shows cytotoxic effects at concentrations far below the CMC [9]. Therefore, we were interested to investigate whether hexadecylphosphocholine (HePC) also causes damage of plasma membranes of tumor cells in sublytic concentrations and if the effects are similar to the effects caused by OcMeG3PC [6, 9, 11]. In the present publication, we discuss the effects of HePC on the plasma membrane of the lymphomic Raji cell line and of the adherent KB, MDA-MB231 and 3T3 cell lines in comparison to the effects of OcMeG3PC.

[1] We thank W.D. Kötting and J. Winkler for advice in microscopy.

Materials and Methods

Materials

HePC and OcMeG3PC were synthesized in our laboratory as described in the first chapter in this volume by Eibl and Engel. Stock solutions of both lipids in concentrations of 1 mg/ml were prepared in RPMI 1640 (Gibco, Karlsruhe, FRG), supplemented with 10% fetal calf serum (FCS; Biochrom, Berlin, FRG). After sterile filtration with Minisart SM16534/55 (Sartorius, Göttingen, FRG) the stock solutions were kept at 4 °C. They were stable for at least 10 months.

Cell Lines

Nonadherent lymphomic Raji cells (B-lymphoblastic lymphoma) and the three adherent cell lines 3T3 (Swiss 3T3 murine fibroblasts), MDA (MDA-MB 231, breast carcinoma cells), and KB (oral epidermoid carcinoma cells) were obtained from American Type Culture Collection (Rockville, Md., USA) and used in the experiments.

Preparation of Glass Coverslips

Normal glass coverslips were washed and soaked in concentrated hydrochloric acid for 2 h. After treatment, the coverslips were rinsed with distilled water until neutral pH, and sterilized in an autoclave.

In vitro Experiments

Cells were allowed to settle (Raji) or to grow (3T3, MDA, KB) into dense monolayers in Multiwell-6 Tissue Culture Plates (Becton Dickinson, Heidelberg, FRG) on pretreated glass coverslips. Starting cell density was 1×10^5 cells per well for adherent cells and 4×10^6 cells per well for Raji cells. Cells were incubated with 2 ml RPMI 1640 per well, supplemented with 10% FCS, streptomycin (200 mg/l), penicillin (2×10^5 IU/l) and fresh *l*-glutamine (200 m*M*, Gibco, Karlsruhe, FRG) at 37 °C, in an atmosphere of 5% CO_2, and high humidity. After 2 days, a cell density of 7×10^5 cells per well was reached for adherent cells. The culture medium of 3T3, MDA and KB was changed and the cells were incubated for 24 h with HePC and OcMeG3PC in concentrations ranging from 0.5 up to 40 μg/ml culture medium. These lipid concentrations are sublytic in the presence of 10% FCS.

Fixation and Preparation for Scanning Electron Microscopy (SEM)

Cells were fixed in a solution of 2% glutaraldehyde in 0.1 *M* sodium cacodylate buffer with 0.1 *M* sucrose at pH 7.4. The fixation process was initiated by slowly adding the prewarmed fixative (37 °C) to the original medium in equal volumes. In this way, care was taken to avoid changes in mobile structures [12]. After 5 min the mixture was removed and cells were incubated in pure fixative at room temperature for 3 h. Afterwards, cells were carefully washed four times with 0.1 *M* sodium cacodylate buffer pH 7.4, supplemented with 0.1 *M* sucrose. Cells were dehydrated in a graded series of ethanol (50, 75, 90, and 100%) and acetone, dried from CO_2 in a Critical Point Dryer (Balzers Union, Balzers, Liechtenstein), fixed with double-faced tape on appropriate stubs, covered with gold in a Sputtering Device (Balzers Union, Balzers, Liechtenstein) and examined in a Cambridge Stereoscan 150 (Cambridge Instruments, Nussloch, FRG).

Testing Proliferation and Viability of Cells

Before viability testing, representative samples of attached cells were trypsinized and washed twice with culture medium. Aliquots of cells were stained with trypan blue and counted for viability in a Neubauer hemocytometer using the trypan blue dye exclusion assay as described by Hudson and Hay [13]. Viability of cells before treatment with HePC or OcMeG3PC and viability of the control cells after incubation were higher than 95%. From a series of cells incubated with increasing concentration of agents, the 50% lethal dose (LD_{50}) of HePC or OcMeG3PC was interpolated from data obtained by counting cell numbers (Neubauer hemocytometer) as described before [14].

Results

Cytotoxic Effects

Table 1 summarizes ID_{50} and LD_{50} values of the different cell lines for OcMeG3PC and HePC after 48 h of incubation at 37 °C. The adherent cell line Swiss 3T3, a nondifferentiated transformed cell line can only be affected with relatively high concentrations of HePC ($ID_{50} > 85\ \mu M$, $LD_{50} > 85\ \mu M$) or OcMeG3PC (ID_{50} 21 μM, LD_{50} 41 μM). MDA cells are more sensitive (HePC, ID_{50} 19.1 μM, LD_{50} 41.2 μM), and KB cells are most sensitive (HePC, ID_{50} 1.6 μM, LD_{50} 3.1 μM; OcMeG3PC, ID_{50} 1.1 μM, LD_{50} 1.6 μM). Raji cells fall somewhere close to MDA cells with respect to their sensitivity for HePC (ID_{50} 7.1 μM, LD_{50} 58.8 μM). Note, however, that individual cell lines often show a large variation in their ID_{50}/LD_{50} ratios. For instance, Raji cells are more sensitive than MDA with respect to growth inhibition for HePC (ID_{50} of Raji 7.1 μM and ID_{50} of MDA 19.1 μM, respectively). On the other hand, Raji cells have a higher LD_{50} (58.8 μM) than MDA (41.2 μM) when incubated with HePC. A comparison of HePC and OcMeG3PC shows that KB cells have nearly identical ID_{50} and LD_{50} values for these two agents. The other, less sensitive cell lines show a higher sensitivity to OcMeG3PC than to HePC. This is probably a result of the longer alkyl chain in the case of OcMeG3PC.

Table 1. Influence of HePC and OcMeG3PC on cell proliferation and cell viability

	HePC		OcMeG3PC	
	ID_{50}	LD_{50}	ID_{50}	LD_{50}
Raji	7.1	58.8	13.4	25.8
KB	1.6	3.1	1.1	1.6
MDA	19.1	41.2	n.d.	n.d.
3T3	> 85	> 85	21.0	41.1

Effects on different cell lines after 48 h incubation at 37 °C. The ID_{50} and LD_{50} values for HePC or OcMeG3PC are concentrations [μM] and show a standard deviation of $\pm 10\%$; n.d. = not determined.

1 2 3 4

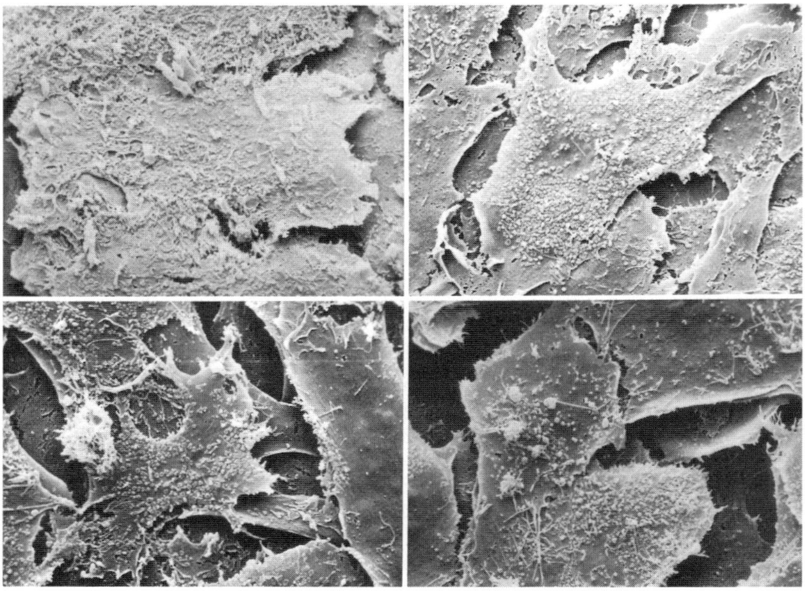

Fig. 1–4. Scanning electron micrographs of 3T3 cells untreated *(1)*, or treated with HePC: *(2)* 10 μg/ml (25 *μM*); *(3)* 20 μg/ml (50 *μM*) and *(4)* 40 μg/ml (100 *μM*) for 24 h. Bar is equivalent to 10 *μm*.

Observations on Cell Morphology

Murine Swiss 3T3 fibroblasts, which are non-neoplastic, have a huge, flat and irregular shape (fig. 1, 5). Their size is about 60 *μm* in longitudinal direction and 30 *μm* in cross direction. They adhere poorly to each other and do not appear to sense their neighbours. These 3T3 control cells spread out into thin lamellae with many knob-, disc- and finger-like extensions on their surfaces. In numerous lamellopodia we can find small holes even in control cells. No evident changes in cell surface or cell shape are observed at 10 μg HePC per ml (25 *μM*) (fig. 2). At HePC concentrations of 20 μg/ml (50 *μM*), cell density decreases indicating growth inhibition and a possible loss of adherence to the substrate (fig. 3). When incubated with 40 μg HePC per ml (100 *μM*) many cells are strongly affected, showing advanced blebbing on their surfaces instead of disc- or finger-like extensions (fig. 4).

After incubation with OcMeG3PC, 3T3 cells show similar effects as described for incubation with HePC: smoothing of the cell surface (10 μg OcMeG3PC per ml, e.g. 19 *μM*, fig. 6), decreased cell density with advanced blebbing (20 μg OcMeG3PC per ml, e.g. 38 *μM*, fig. 7) and severe

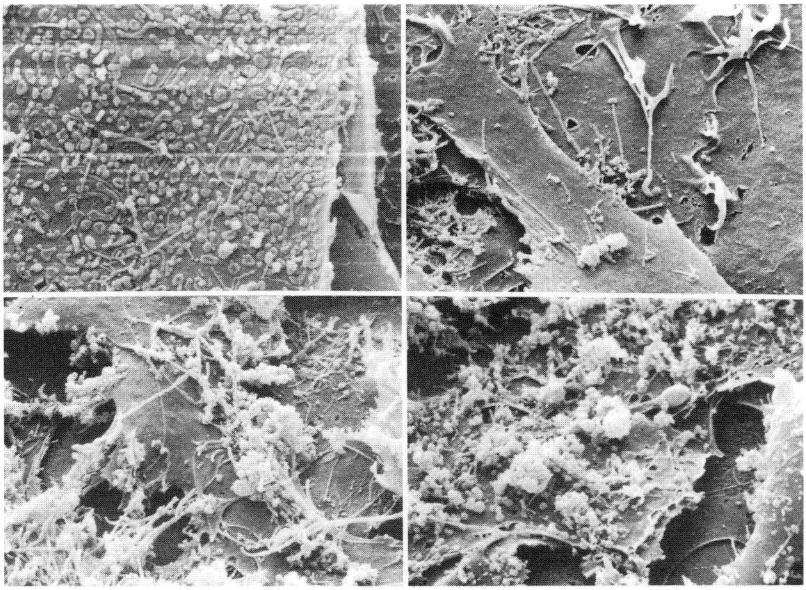

Fig. 5–8. Scanning electron micrographs of 3T3 cells untreated *(5)*, or treated with OcMeG3PC: *(6)* 10 µg/ml (19 *µM*); *(7)* 20 µg/ml (38 *µM*); *(8)* 40 µg/ml (76 *µM*) for 24 h. Bar is equivalent to 4 *µm*.

blebbing (40 µg OcMeG3PC per ml, e.g. 76 *µM*, fig. 8). In spite of other excessive cell surface alterations even at higher concentrations, namely 80 µg OcMeG3PC (152 *µM*) or 80 µg HePC per ml (200 *µM*, not shown here), we still observe cells in original shape and size. In no case did we observe enhanced pore formation in the plasma membrane compared with the control cells.

Figure 9 shows MDA control cells with their typical flattened spindle form (60–70 *µm* in length) and many small multishaped extensions on the cell surface. There seems to be a firm adherence to the substrate and an extensive overlapping of cells at many places of direct intercellular contact. Cells are spread out in monolayers and cell density is quite high. After 24 h of incubation with 10 µg HePC per ml (25 *µM*) MDA cells are still looking quite normal; however, note the decreasing number of small multishaped extensions (fig. 10). Treatment with 20 µg HePC (fig. 11, 50 *µM*) generates a sequence of serious cell alterations: in addition to growth inhibition, the greater part of cells no longer adheres to the substrate, so that cell density decreases very strongly. Similarly, cells lose contact to each other and very

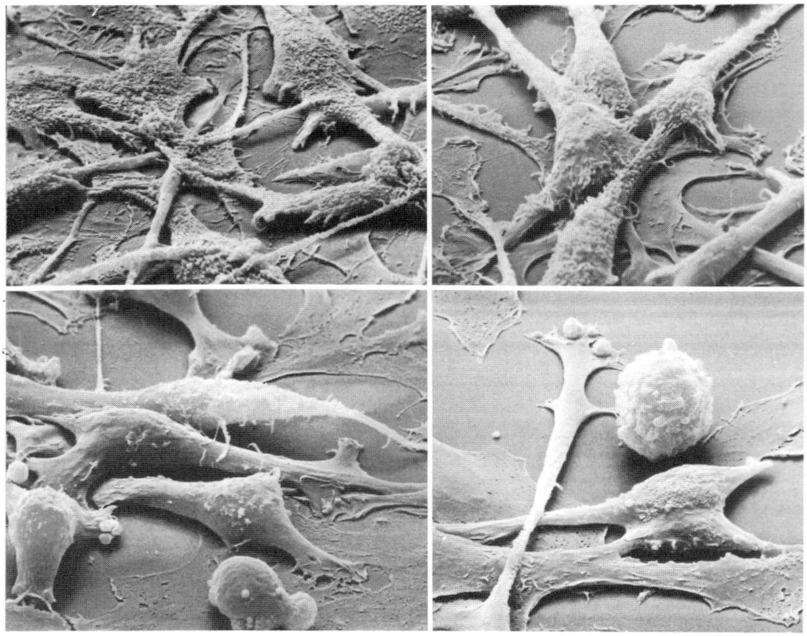

9 10 11 12

Fig. 9–12. Scanning electron micrographs of MDA cells untreated *(9)*, or treated with HePC: *(10)* 10 μg/ml (25 μ*M*); *(11)* 20 μg/ml (50 μ*M*) and *(12)* 40 μg/ml (100 μ*M*) for 24 h. Bar is equivalent to 10 μ*m*.

often are more rounded, which is thought to be the result of disappearing stress fibers [15]. At this time, remaining cells differ in size and shape and most of them have a smooth surface or show extensive blebbing. Although cells at this stage are thought no longer to be alive, holes or other pertubations of the plasma membrane can only be observed in totally damaged cells and cell debris (fig. 12).

Untreated Raji cells do not show a high cell density since this cell line is nonadherent. In this case cells were only able to adhere on glass coverslips because these are specially pretreated with hydrochloric acid (see 'Materials and Methods'). The control cells are round with a diameter of 7 μ*m*. They are sticking together and are in contact with each other. Normal, untreated cells show an intact cover of finger-like extensions, so-called microvilli, with a length of about 1–2 μ*m* and a diameter of 0.1–0.2 μ*m* (fig. 13, 16). With increasing HePC concentration we observe a decreasing cell density, which is probably not only the result of growth inhibition but also of loss of adherence between the cells or to the substrate. At HePC concentrations of 10 μg/ml (25 μ*M*) cells still maintain

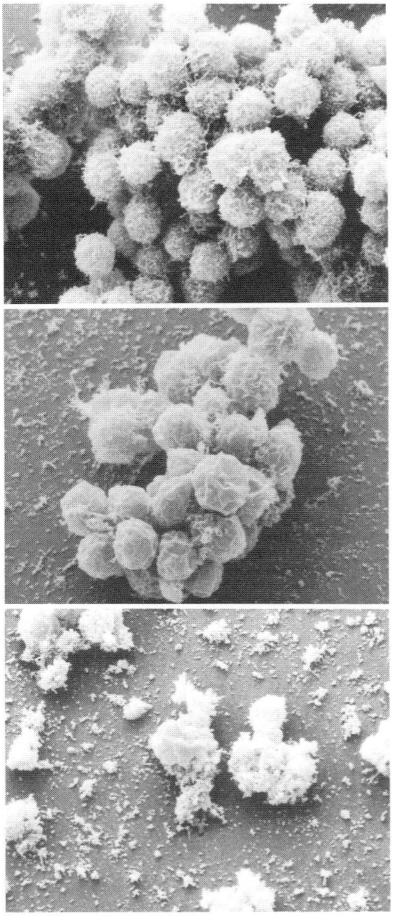

13

14

15

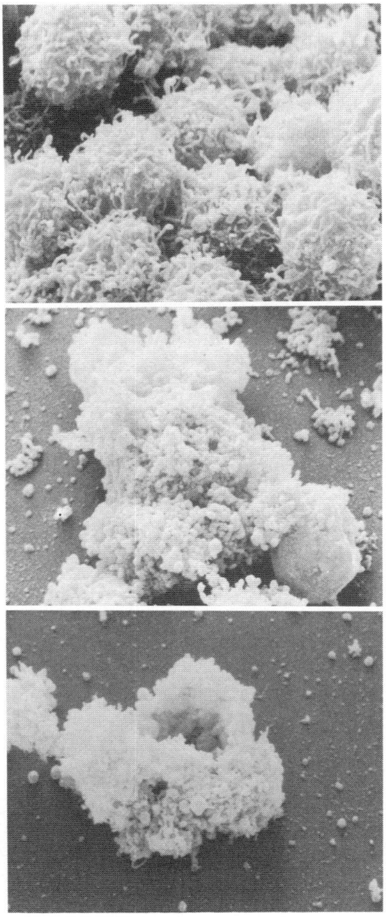

16

17

18

Fig. 13–15. Scanning electron micrographs of Raji cells untreated *(13)*, or treated with HePC: *(14)* 10 μg/ml (25 *μM*); *(15)* 20 μg/ml (50 *μM*) for 24 h. Bar is equivalent to 10 *μm*.

Fig. 16–18. Scanning electron micrographs of Raji cells untreated *(16)*, or treated with OcMeG3PC: *(17)* 10 μg/ml (19 *μM*) and *(18)* 20 μg/ml (38 *μM*) for 24 h. Bar is equivalent to 4 *μm*.

their integrity but show definite signs of succumbing (fig. 14). Their surfaces are almost bare of microvilli and already smoothing out with large foldings. Treatment with even higher concentrations of HePC (20 μg/ml; 50 *μM*) results in complete cell destruction. Besides strongly affected cells with many blebs on the surfaces, only cell fragments can be observed on the coverslips (fig. 15).

Similar effects can be seen under treatment of Raji cells with OcMeG3PC at lower concentrations. At a dose of 10 μg/ml (19 μM), OcMeG3PC destroys the Raji cells completely. As shown in figure 17, cells have already lost their integrity and seem to consist of a mass of blebs. Figure 18 shows a dead cell after incubation with 20 μg OcMeG3PC per ml (38 μM) with an area of complete destruction, exposing the interior of the cell.

KB cells are adherent round cells with a diameter of about 14 μm. These cells cling together and attach to the substrate to a great extent, so that their cell density is extremely high (fig. 19, 22). Normal KB cells show an intact and very dense cover of microvilli, which are about 0.5–2 μm long. Already at concentrations of 0.5 μg HePC per ml (1.25 μM) this cover of microvilli on the cell surface seems to be thinning – possibly a first sign of being affected by HePC (not shown here). Incubation with 1 μg HePC per ml (2.4 μM) leads to a dramatically reduced number of cells and to severe alterations on the cell surface, namely appearance of huge blebs corresponding to a decreasing number of intact microvilli and intercellular contacts (fig. 20). Higher concentrations of HePC cause complete cell destruction (fig. 21).

Again we can observe similar surface alterations after incubation with OcMeG3PC. In contrast to the Raji cells, where OcMeG3PC has a stronger effect than HePC, the KB cells are less sensitive to OcMeG3PC than to HePC. This result is at variance with the data from table 1, where HePC and OcMeG3PC show almost identical ID_{50} values for KB cells. After treatment with 1 μg OcMeG3PC per ml (1.9 μM) KB cells do not show any difference when compared with untreated control cells (fig. 23). At concentrations of 10 μg OcMeG3PC per ml (19 μM), the coverslip is covered with dying cells as well as with dead cells, showing typical blebbing and cell debris (fig. 24).

Discussion and Conclusion

Scanning electron microscopy is a widely used technique to study morphological changes on surfaces of normal or tumor cells, respectively [9, 16]. In our case, results from observations of Raji, KB, MDA, and 3T3 cells treated with HePC or OcMeG3PC show a close correlation with the corresponding data obtained by testing growth inhibition and cytotoxicity. Although SEM observations are descriptive and only show visible alterations on cell surfaces generated by conversions inside the cell, these morphological changes were seen with SEM much earlier and at lower concentrations of the agents than the cytotoxic effects observed by

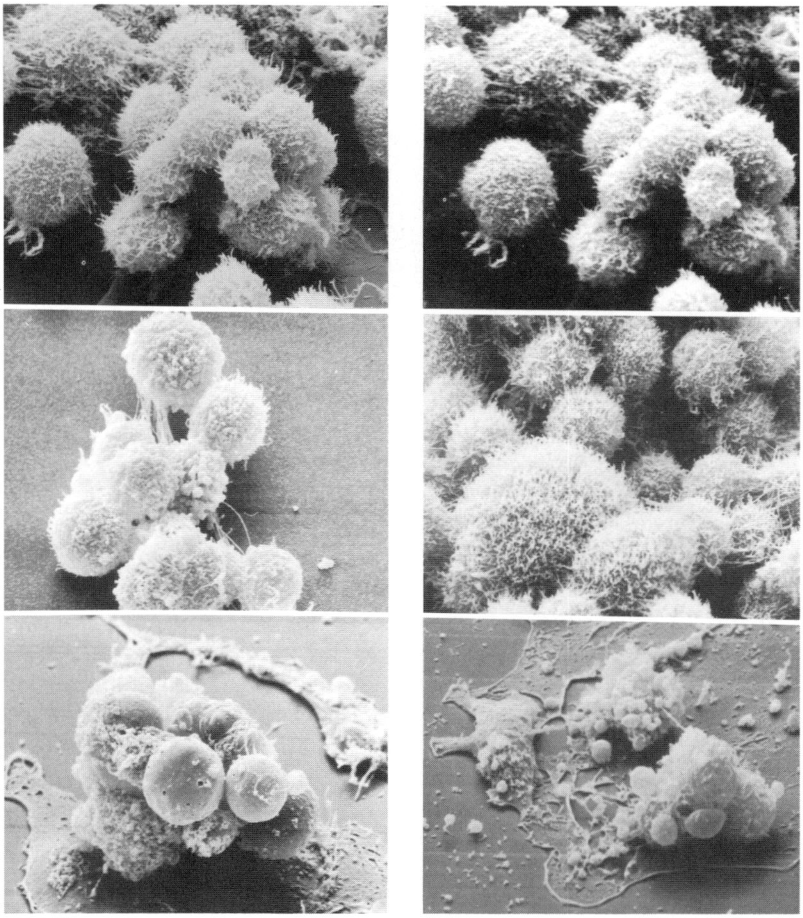

19 22 20 23 21 24

Fig. 19–21. Scanning electron micrographs of KB cells untreated *(19)*, or treated with HePC: *(20)* 1 μg/ml (2.5 μ*M*) and *(21)* 10 μg/ml (25 μ*M*) for 24 h. Bar is equivalent to 10 μ*m*.

Fig. 22–24. Scanning electron micrographs of KB cells untreated *(22)*, or treated with OcMeG3PC: *(23)* 1 μg/ml (1.9 μ*M*) and *(24)* 10 μg/ml (10 μ*M*) for 24 h. Bar is equivalent to 10 μ*m*.

using the trypan blue dye exclusion assay. For instance, in observations with SEM, Raji cells already show severe membrane alterations after 24 h of incubation in relatively low concentrations of HePC or OcMeG3PC, whereas for growth inhibition tested by cell counting, higher amounts are needed for 48 h of incubation (table 1).

The tested cell lines are not affected to the same extent. Two of them, namely 3T3 and MDA, are more resistant to OcMeG3PC and HePC than

Raji or KB. However, the observed effects are similar for both compounds tested, when compared on the same cell line. This indicates that the mechanism of interaction with these cell lines for both compounds may be similar.

Besides the well-documented cytotoxic effects [1–8], OcMeG3PC and HePC also have strong lytic properties with lytic concentrations of 1 and 8 μM, respectively. In the presence of fetal calf serum cells are protected from lysis of OcMeG3PC and HePC over a wide range of concentrations, because of binding to serum proteins [unpubl. data]. This means that under conditions of our incubation system, the chosen concentration of OcMeG3PC or HePC in the range of 1–80 μM is below the lytic concentration of the compounds in the presence of the used fetal calf serum. Only at concentrations above 100 μM in the presence of 10% fetal calf serum, could the effects be explained by cell lysis. Therefore, at concentrations used in table 1 more specific interactions are involved in the cytotoxic action of HePC and OcMeG3PC than unspecific cell lysis.

In agreement to these observations it was reported that HL60 cells and K562 cells which were exposed to 1-acyl-*sn*-glycero-3-phosphocholine (LPC), a compound without any cytotoxic or antineoplastic effects, show membrane alterations which differ strongly from that seen with OcMeG3PC [9]. LPC, which can be metabolized very quickly causes membrane lesions such as small but diffuse punctures at a concentration of 3 μM. In contrast, the same cell lines show blebbing, formation of large and well-defined holes and generalized porosity after treatment with OcMeG3PC at 2.5 μM [9]. Other structurally related compounds such as (thio-alkyl)-lysophosphocholines or completely different compounds such as 4-aminomethyl-1-[2,3-(di-*n*-decyloxy)-*n*-propyl]-4-phenylpiperidine (CP-46,665) produced similar effects [11, 17]. After treatment with these compounds cytoplasmic processes, described as numerous blebs, ruffles and filopodia, vanished. Instead of this, many round defects, resembling punched holes, could be observed in the plasma membrane [17]. Furthermore, many cells lost their adherence properties [11].

In none of the cell lines used in this study, did we observe pore formation in the plasma membrane after incubation with OcMeG3PC or HePC, respectively. However, in SEM we could see similar effects for both OcMeG3PC and HePC: (i) a decreased cell density, which is probably a result of growth inhibition and of loss of adherence between cells or to the substrate (this makes it impossible to evaluate SEM micrographs quantitatively); (ii) smoothing of the cell surface, meaning loss of microvilli and other disc-like protrusions; (iii) blebbing, and (iv) changes in cell shape. The extent of these effects were, of course, dependent on the sensitivity of the respective cell line for the agent in question.

Differences between our observations and those described by others may be due to the use of other cell lines or to the different methods of cell preparation. Although the mechanism by which 1-O-alkyl-glycero-3-phosphocholines or alkylphosphocholines destroy cells is not completely understood, we suggest that the shown effects are not specific for these lipids but are in general typical for cell death as a consequence of cytoskeletal damage [15] or of an imbalance of Ca^{2+} content of the cells [18].

In our opinion the observed changes in the morphology of the tested cell lines probably are the last stage in a long sequence of many different and severe changes in intracellular processes. These changes occur much earlier and perhaps already at lower doses of OcMeG3PC, HePC or other related compounds, than at the concentrations used in our experiments.

References

1 Munder PG, Berchtold R, Palke W: Tumorzellzerstörung durch Alkyllysophospholipide: Beziehung zwischen chemischer Struktur and biologischer Aktivität; in Unger C, Eibl H, Nagel GA (eds.): Die Zellmembran als Angriffspunkt der Tumortherapie. Aktuelle Onkologie München, Zuckschwerdt Verlag, 1987, vol, 34, pp 37–55.

2 Storme GA, Bruyneel EA, Schallier DC, Bolscher JG, Berdel WE, Mareel MM: Effect of lipid derivatives on invasion in vitro and on surface glycoproteins of three rodent cell types. Lipids 1987;22:847–850.

3 Hochhuth C, Berkovic D, Eibl H, Unger C, Doenecke D: Effects of antineoplastic phospholipids on parameters of cell differentiation in U937 cells. J Cancer Res Clin Oncol 1990;116:459–466.

4 Schick HD, Berdel WE, Fromm M, Fink U, Jehn U, Ulm K, Reichert A, Eibl H, Unger C, Rastetter J: Cytotoxic effects of ether lipids and derivatives in human nonneoplastic bone marrow cells and leukemic cells in vitro. Lipids 1987;22:904–910.

5 Fleer EAM, Kim DJ, Nagel GA, Eibl H, Unger C: Cytotoxic activity of lysophosphatidylcholine analogs on human lymphoma Raji cells. Oncology 1990;13:295–300.

6 Eibl H, Unger C: Hexadecylphosphocholine – a new and selective antitumor drug. Cancer Treat Rev 1990;17:233–242.

7 Muschiol C, Berger MR, Schuler B, Scherf HR, Garzon FT, Zeller WJ, Unger C, Eibl H, Schmähl D: Alkylphosphocholines: Toxicity and anticancer properties. Lipids 1987;22:930–934.

8 Hilgard P, Stekar J, Vögeli H, Engel J, Schuhmacher W, Eibl H, Unger C, Berger MR; Characterization of the antitumor activity of hexadecylphosphocholine (D 18506). Eur J Cancer Res Clin Oncol 1988;24:1457–1461.

9 Noseda A, White JG, Godwin PL, Jerome WG, Modest EJ: Membrane damage in leukemic cells induced by ether and ester lipids: An electron microscopic study. Exp Mol Pathol 1989;50:69–83.

10 Snyder F, Record M, Smith Z, Blank ML, Hoffmann DR: Selective cytotoxic action of ether lipid analogs of PAF; mechanistic studies related to their metabolism, subcellular localization, and effects on cellular transport systems; in Unger C, Eibl H, Nagel GA (eds): Die Zellmembran als Angriffspunkt der Tumortherapie. Aktuelle Onkologie. Müchen, Zuckschwerdt Verlag, 1987, vol 34, pp 19–26.

11 Berdel WE, Schick HD, Fink U, Reichert A, Ulm K, Rastetter J: Cytotoxicity of the alkyl-linked lipoidal amine 4-aminomethyl-1-[2,3-(di-*n*-decyloxy)-*n*-propyl]-4-phenylpiperidine (CP-46,665) in human tumors and leukemias. Cancer Res 1985;45:1206–1213.

12 Arro E, Collins VP, Brunk UT: High resolution SEM of cultured cells: Preparatory procedures. Scanning Electron Microsc 1981;II:159–168.

13 Hudson L, Hay FC (eds): Practical Immunology. Oxford, Blackwell Scientific Publications, 1976, pp 29–32.

14 Fleer EAM, Unger C, Kim DJ, Eibl H: Metabolism of ether phospholipids and analogs in neoplastic cells. Lipids 1987;22:856–861.

15 Darnell J, Lodish H, Baltimore D (eds): The Cytoskeleton and Cellular Movements: Actin-Myosin; in: Molecular Cell Biology. New York, Scientific American Books, 1986, pp 815–858.

16 Maistry L, Robinson KM, Evers P, Munder PG, Andreesen R: Morphological effects of an antitumor agent on human esophageal carcinoma cells in vitro. Scanning Electron Microsc 1980;III:109–114.

17 Berdel WE, Fromm M, Fink U, Palke W, Bicker U, Reichert A, Rastetter J: Cytotoxicity of thioether-lysophospholipids in leukaemias and tumors of human origin. Cancer Res 1983;43:5538–5543.

18 Lemasters JJ, DiGuiseppi J, Nieminen AL, Herman B: Blebbing, free Ca^{2+} and mitochondrial membrane potential preceding cell death in hepatocytes. Nature 1987;325:78–81.

Petra Kaufmann-Kolle, Max-Planck-Institut für biophysikalische Chemie, Abteilung Membranbiophysik, Am Fassberg, D-W–3400 Göttingen (FRG)

Eibl H, Hilgard P, Unger C (eds): Alkylphosphocholines: New Drugs in Cancer Therapy. Prog Exp Tumor Res. Basel, Karger, 1992, vol 34, pp 59–68

Effects of Hexadecylphosphocholine on Cellular Function

Dinko Berkovic[a], *Eduard A.M. Fleer*[a], *H. Eibl*[b], *C. Unger*[a]

[a]Medizinische Universitätsklinik, Abteilung Hämatologie/Onkologie, and
[b]Max-Planck-Institut für biophysikalische Chemie, Göttingen, FRG

Introduction

Hexadecylphosphocholine (HePC) is a representative of a new group of antineoplastic agents. Initially, HePC was developed on the basis of structure variation studies of ether phospholipids with respect to their antiproliferative characteristics [1]. Though quite a few data on the cytotoxic or cytostatic activity of HePC exist [2], little is known so far about the molecular mechanism by which HePC exerts its biological effects. In the following chapter, we would like to update information gathered in the past 2 years on changes of cellular function caused by HePC.

Growth Inhibitory and Cytotoxic Effects

The antiproliferative effects of HePC were tested in a variety of tumor cell lines in vitro. Table 1 summarizes data from our laboratory on cell proliferation parameters in seven different cell lines. Apparently, different cell lines show a broad variety of cellular susceptibility towards HePC. The epidermoid cancer cell line KB and the promyelocytic cell line HL60 are most sensitive while K562, a erythroleukemia cell line and the mouse fibroblast line Swiss 3T3 are almost resistant.

For the ether lipid prototype ET-18-OCH_3 similar patterns of cellular sensitivity have been described [3–5]. This may reflect a similar or even identical mechanism of action of both compounds.

The interesting question arises what reasons there may be for such differences in cellular susceptibility. It was demonstrated by Unger [2] and Fleer et al. [6] that neither the cellular uptake of HePC nor the phospholipid and cholesterol content of the membranes can explain the variation of

Table 1. Cell proliferation parameters of seven different cell lines

Cell line	ID_{50} (a)	ID_{50} (b)	LD_{50}
KB	0.6 ± 0.2	2.2 ± 1.2	1.2 ± 0.9
MDA MB 231	7.8 ± 1.6	20.6 ± 5.2	16.8 ± 2.6
HL 60	2.6 ± 1.8	2.2 ± 1.3	2.1 ± 1.0
K 562	23.1 ± 1.7	27.2 ± 1.8	>35
U 937	2.4 ± 1.2	3.2 ± 0.2	7.1 ± 0.9
Raji	2.9 ± 1.7	6.9 ± 2.9	24.0 ± 2.4
Swiss 3T3	>35.0	n.d.	>35

Cells were incubated for 48 h at 37 °C with increasing concentrations of HePC. ID_{50} values were determined by cell count (a) or ^{3}H-thymidine incorporation (b). LD_{50} values were calculated from trypan blue dye exclusion assays. Data are mean values ± SD of 3–6 independent experiments.

cell sensitivity. Though, at least for ET-18-OCH_3, it has been shown that cell specific constitutive ether lipid content in the membranes may be of importance insofar as cells containing higher amounts of etherphospholipids are more sensitive [7].

On the other hand, cellular sensitivity towards ET-18-OCH_3 also seems to depend on the state of cell differentiation [8] explaining perhaps the fact that mouse fibroblasts and normal fully differentiated tissues in animal experiments are not affected by HePC.

Effects of HePC on the Membrane Transport of Cellular Nutrients

As HePC is predominantly incorporated in the cellular lipid compartments it was interesting to investigate possible interactions of HePC with physiological cell membrane functions. One of the membrane functions tested in our laboratory was the uptake of some essential cell nutrients like myo-inositol, choline, methionine and glucose in the human Burkitt lymphoma cell line Raji [9]. HePC caused a marked, time dependent inhibition of ^{3}H-myo-inositol uptake in these cells (fig. 1). Already after 10 min of exposure to HePC a 15% decrease from control uptake could be observed. Extension of HePC treatment resulted in a 50% decrease of uptake after 2 h, and an almost 90% uptake inhibition after 24 h of preincubation. The effects of HePC on ^{14}C-desoxyglucose, ^{14}C-choline and ^{14}C-*L*-methionine uptake were smaller. After 2 h of HePC exposure for example the incorporation of ^{14}C-desoxyglucose was inhibited by 35%, the uptake of ^{14}C-choline by 26% and the uptake of ^{14}C-*L*-methionine by only

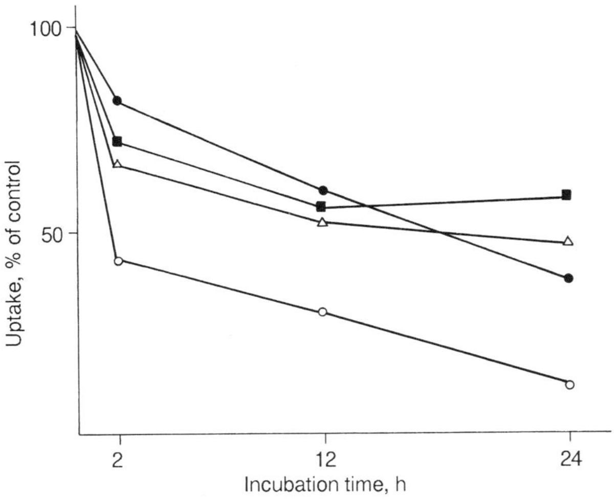

Fig. 1. Inhibitory effect of HePC (10 μg/ml) on the cellular uptake of ^{3}H-myo-inositol (○), ^{14}C-choline (■), ^{14}C-*L*-methionine (●) and ^{14}C-desoxyglucose (▲) in Raji cells. Uptake was measured over 10 min after pretreatment of cells with HePC for indicated time periods. Data show mean values of four independent experiments. SD $\pm$ 6%.

17% (fig. 1). These variations of uptake inhibition and of uptake inhibition kinetics suggest certain differences of sensitivity of these cellular transport systems towards HePC.

To show that uptake inhibition correlates with the antiproliferative effects of HePC, cells were pretreated with equimolar concentrations of ester lysophosphatidylcholine (ES-16-OH) and hexadecyl-N,N,N-trimethylhexanolamine ($HePC_6$), a compound with almost identical physicochemical properties but less antiproliferative potential (fig. 2). As expected $HePC_6$ reduced ^{3}H-myo-inositol uptake to a lesser degree than HePC after 2 and 24 h of incubation. Surprisingly, ES-16-OH showed an inhibitory effect after 2 h of pretreatment, whereas after 24 h no difference to control could be detected. This observation may be explained by reacylation of ES-16-OH in the sn-2 position, converting the lysolipid to a diacyl-phospholipid. In contrast, HePC and $HePC_6$ cannot be metabolized by the cells in this short period of time and remain in their active form. The exact mechanism, however, by which the uptake of these nutrients is suppressed is still unknown. HePC might directly or indirectly interfere with specific carrier proteins as postulated for the Na^{+},K^{+}-ATPase [10] leading eventually to a depletion of essential nutrients, growth retardation and cell death.

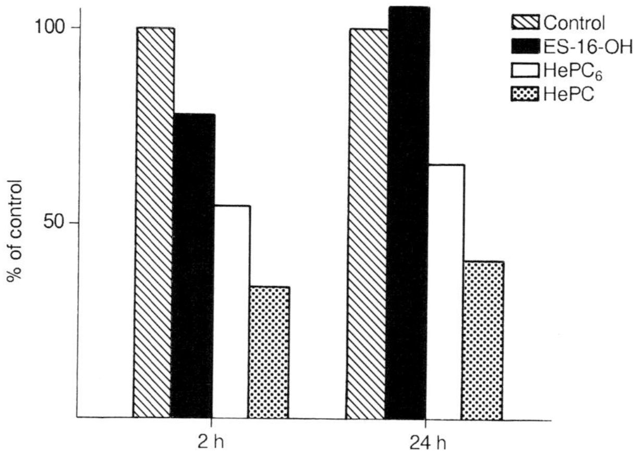

Fig. 2. Uptake of ^{3}H-myo-inositol in Raji cells. Uptake was determined over 10 min after pretreatment of cells for 2 h (left column) and 24 h (right column) with HePC (10 μg/ml), $HePC_6$ (11.6 μg/ml) and ES-16-OH (12.8 μg/ml). Data represent mean of three different experiments with a SD $\pm$ 9%. These experiments were performed by W. Menzel (University Hospital, Göttingen, FRG).

Effects of HePC on Phospholipid Synthesis

In a following series of experiments it was important to prove whether the inhibitory activity of HePC on choline and myo-inositol uptake substantially affects the synthesis and metabolism of choline or inositol containing phospholipids.

We therefore investigated the influence of HePC on phospholipid metabolism in Raji cells and in the human epidermoid cancer cell line KB, which is very sensitive to the antiproliferative effects of HePC. Indeed, we noticed a decrease of ^{14}C-choline incorporation in phosphatidylcholine after 24 h of HePC treatment up to 30% in Raji and up to 40% in KB cells (table 2). In both cell lines most of the label was found in phosphatidylcholine, but in the presence of HePC probably a slight shift of label from phosphatidylcholine to sphingomyelin was observed. Lysophosphatidylcholine levels remained almost constant. In order to directly label membrane phosphatidylcholine, cells were incubated with lysophosphatidyl-[^{3}H]-choline in the presence of HePC. This time, the transfer of label to sphingomyelin was even more pronounced, suggesting an enhanced phospholipase C-like activity in these cells rather than a stimulation of sphingomyelin synthesis via CDP-choline.

Table 2. Incorporation of ^{14}C-choline, ^{3}H-lysophosphatidylcholine, ^{14}C-arachidonic acid and ^{3}H-myo-inositol in the membrane lipids of 2×10^6 Raji or 5×10^5 KB cells

	HePC µg/ml	LPC	SM	PC			
^{14}C-choline							
RAJI	0	0.6	3.4	96.0			
	2	0.5	4.7	81.7			
	4	0.6	4.5	61.8			
KB	0	1.3	5.7	91.0			
	0.5	0.9	4.2	66.7			
	1	0.8	5.5	52.8			
1-Palmitoyl-lyso-phosphatidyl[^{3}H]-choline							
RAJI	0	2.5	9.6	87.1			
	2	4.3	17.2	76.9			
	4	5.6	21.5	73.7			
KB	0	1.7	11.0	83.8			
	0.5	3.2	16.9	77.7			
	1	2.5	16.7	75.5			
^{14}C-arachidonic acid					PS/PI	PE	NL
RAJI	0	1.0	0.0	19.9	22.2	45.5	8.8
	2	1.4	0.0	21.5	24.0	39.1	11.5
	4	1.9	0.0	22.2	23.2	35.4	12.9
KB	0	1.2	3.0	43.3	0.0	17.9	30.4
	0.5	1.9	2.1	45.9	0.0	19.5	45.1
	1	1.4	2.2	36.9	0.0	15.4	60.4
^{3}H-myo-inositol		PIP_2	PIP	PI			
RAJI	0	1.8	3.9	92.3			
	2	1.8	6.0	91.3			
	4	5.4	4.6	88.4			
KB	0	4.9	12.3	62.2			
	0.5	4.9	11.2	60.7			
	1	5.4	9.0	64.6			

Incubations were performed in presence of indicated HePC concentrations for 24 h at 37 °C. Data show mean values of three independent experiments expressed as percentage of radiolabel incorporation in untreated control cultures.

Looking at the synthesis of phosphatidylinositols the strong inhibition of ^{3}H-myo-inositol uptake was not reflected by a decreased incorporation into phosphatidylinositols (table 2, bottom). After 24 h of HePC treatment with escalating doses, no changes of myo-inositol incorporation were detectable in either cell line. Analysis of phosphatidylinositols by thin-layer chromatography revealed a dose-dependent accumulation of label in phosphatidylinositolbisphosphate with a concomitant loss in phosphatidylinositol. In contrast to the experiments with choline one would assume an inhibitory effect of HePC on phospholipase C, in this case a phosphatidylinositol specific phospholipase C.

Finally, cells were also loaded with ^{14}C-arachidonic acid in the presence of HePC. In KB cells an increase of total ^{14}C-arachidonic acid incorporation of about 20% over control values was noticed. In Raji cells there was no difference to control cultures. As shown in table 2, a dose-dependent redistribution of label from phosphatidylcholine in favour of the neutral lipid fraction took place. In Raji cells a similar accumulation of arachidonic acid in neutral lipids was seen. On the other hand, instead of phosphatidylcholine, phosphatidylethanolamine predominantly seemed to be the source of label.

Taken together these data indicate that HePC influences phospholipid metabolism at several sites: It inhibits phosphatidylcholine synthesis from choline and may enhance phosphatidylcholine degradation by a phospholipase C activity. Furthermore, it seems to either decrease reacylation or potentiate a phospholipase A_2-mediated deacylation of phosphatidylcholine in KB and phosphatidylethanolamine in Raji cells. These results are consistent with observations made by others for ET-18-OCH_3. Herrmann [3] showed a reduced incorporation of oleic acid in phosphatidylcholine and an enhanced incorporation in triacylglycerols. Prelabeled phosphatidylcholine was deacylated arguing in favour of a phospholipase A_2 enhancement. Later on, Herrmann and Neumann [11] described a direct acyltransferase inhibition by ET-18-OCH_3 in a cell-free system. Vogler et al. [12] showed a reduction of choline incorporation and an inhibition of lysophosphatidylcholine reacylation. Modolell et al. [13] also observed a decreased formation of phosphatidylcholine from ^{14}C-choline and a redistribution of oleic and linoleic acid into neutral lipids. Additionally, at least in Raji cells, HePC also seems to have an inhibitory effect on the phosphatidylinositol specific phospholipase C as indicated by an accumulation of phosphatidylinositolbisphosphate. In support of this are the results of Seewald et al. [14] who showed reduced inositoltrisphosphate levels after PDGF stimulation in ET-18-OCH_3 treated Swiss 3T3 cells. However, it still remains unclear whether these changes in phospholipid metabolism are crucial for the growth inhibition induced by HePC and ether lipids or merely epiphenomena of rather unspecific membrane perturbing effects.

Protein Kinase C (PKC), Na^+,K^+-ATPase and Sodium Pump

Besides those enzymes involved in phospholipid metabolism two other enzymes have been identified as potential targets of ether phospholipid and HePC action. Membrane-associated enzymes as PKC require a close protein-lipid contact for activation and are regulated by different phospholipids. PKC for instance is stimulated by ester lysophosphatidylcholine in low concentrations whereas in high concentrations it is inhibited [15]. Ether lipids like ET-18-OCH_3, 1-0-hexadecyl-2-0-methyl-glycerol and the thioetherlipid BM.41440 possess PKC inhibitory activities [14–18]. Recently, Zheng et al. showed that HePC similar to a variety of ether phospholipids is a potent inhibitor of PKC and Na^+,K^+-ATPase in a cell-free system and of the sodium pump in intact HL60 cells [10].

Apparently, HePC and ether lipids reduced Na^+,K^+-ATPase activity from brain synaptosomal membranes competitively with respect to Na^+ and noncompetitively with respect to K^+. It is thought that these compounds ‘dilute’ the phospholipid environment of the enzyme, because the inhibitory effect can be overcome by increasing the amount of membrane in this assay system.

Differentiation

As mentioned before, the susceptibility of cells towards ether lipids may depend on their state of differentiation. Surprisingly, ether lipids by themselves seem to possess cell-differentiating activity [19–23]. Though this seems a little contradictory as in some cases phorbol ester induced differentiation could be suppressed by ether phospholipids [18]. Hilgard et al. [22] described granulocyte-like morphological changes in HL60 and M1-T22 cells by subcytotoxic doses of HePC. More interestingly, histological sections of dimethylbenzanthracene-induced rat mammary tumors after treatment of animals with HePC showed a decrease in mitotic rates and a reduction of tumor mass followed by an increase of epithelial differentiation.

A well-established model for cellular differentiation is the human promonocytic leukemia cell line U937. These cells differentiate towards a monocyte-like phenotype upon addition of phorbolester (TPA) and a variety of other substances. We showed in our laboratory that HePC and

Table 3. Changes of surface antigen expression and tetrazolium reduction in U937 cells after 48 h of exposure to TPA, HePC, ET-18-OCH_3, $HePC_6$ and ES-16 OH

Incubation	LeuM3 CD14	Mo2 CD14	Leu3 CD4	Mol CD11	Transferrin receptor	Formazan positive cells %
Control	4 ± 3	28 ± 8	304 ± 21	360 ± 31	268 ± 5	1.6 ± 0.6
TPA 50 ng/ml	6 ± 2	51 ± 9	46 ± 11	463 ± 29	137 ± 11	7.0 ± 1.1
HePC 10 μg/ml	31 ± 6	111 ± 18	218 ± 28	397 ± 20	173 ± 9	17.0 ± 1.3
ET-18-OCH_3 5 μg/ml	47 ± 11	127 ± 10	217 ± 30	385 ± 37	39 ± 13	13.8 ± 1.6
$HePC_6$ 10 μg/ml	5 ± 2	3 ± 1	316 ± 17	352 ± 10	349 ± 9	1.0 ± 0.6
ES-16-OH 10 μg/ml	0	21 ± 5	333 ± 17	361 ± 17	344 ± 13	0.5 ± 0.2

Antigen expression was determined by flow cytometry analysis after immunofluorescence staining with monoclonal antibodies. Tetrazolium reduction capacity was determined by incubation of cultures with 0.2% nitroblue tetrazolium for 45 min. Percentage of cells containing formazan deposits was counted in a Neubauer chamber. Data represents mean fluorescence intensities and percent positive cells, respectively $\pm$SD of three independent experiments.

ET-18-OCH_3 induce several parameters of differentiation in U937 similar to TPA [23]. First of all, these compounds lead to a growth retardation of cells after 24 h of incubation. After 48 h, cell surface antigen expression changes resembling a more mature monocytic pattern (table 3). Typical markers for early stages of the myeloid lineage like transferrin receptors and CD4 antigen disappear from the membrane surface. The expression of CD14 and CD11 antigens, usually detectable on fully differentiated monocytes is enhanced. Though there are some differences in respect to the quantity of antigen levels between TPA, HePC and ET-18-OCH_3, in general all three compounds seem to induce alterations in antigen expression in the same mode. Further evidence for a functional maturation of U937 cells is provided by their increased ability to reduce nitroblue tetrazolium to formazan, a characteristic of phagocytic cells like macrophages, monocytes or granulocytes (table 3, right column). Obviously, induction of differentiation by HePC and ET-18-OCH_3 and cytotoxicity seem to be closely linked phenomena because ES-16-OH and $HePC_6$, two other lipids with almost identical physicochemical properties lacking antiproliferative activity, do not show any effect on differentiation parameters in U937 cells.

Conclusions

Hexadecylphosphocholine is a substance with pleiotropic effects in cell culture. At the membrane level it inhibits specific uptake systems for essential cellular nutrients like myo-inositol or choline. HePC modulates the metabolism of cell membrane lipids at several sites causing for instance a decreased synthesis and possibly an enhanced degradation of phosphatidylcholine. In addition, HePC inhibits two important enzymes of cellular signal transduction and homeostasis, the protein kinase C and the Na^+,K^+-ATPase. To date there is no decisive proof, however, that one of these above-mentioned HePC effects on various cellular functions ultimately leads to a proliferation stop. Perhaps HePC-induced cell death results from a summation of different effects. Another intriguing possibility is that an interference of HePC with certain biochemical pathways does not necessarily lead to cell death but rather to cellular differentiation. Presently, it is unclear whether HePC-induced maturation of U937 or HL60 cells is just part of a sequence of phenotypical changes ending inevitably in cellular destruction or a cellular phenomenon by itself. Nevertheless, especially in vivo differentiation of neoplastic cells may play the predominant role in HePC-mediated antitumoral activity. On the other hand, the most important target structure for HePC may not yet have been identified, thus leaving a broad field for molecular and clinical investigation for the near future.

References

1 Eibl H, Unger C: Hexadecylphosphocholine, a new and selective antitumor drug. Cancer Treat Rev 1990;17:233–242.

2 Unger C: Alkylphosphocholine und Analoga: Entwicklung einer neuen Substanzgruppe mit antineoplastischer Wirkung. Stuttgart, Thieme, 1989.

3 Herrmann DBJ: Changes in cellular lipid synthesis of normal and neoplastic cells during cytolysis induced by alkyl lysophospholipid analogues. J Natl Cancer Inst 1985;75:423–430.

4 Tidwell T, Guzman G, Vogler WR: The effect of alkyl-lysophospholipids on leukemic cell lines. I. Differential action on two human leukemic cells lines, HL60 and K562. Blood 1981;57:794–797.

5 Unger C, Eibl H, Kim DJ, Fleer EAM, Bartsch HH, Nagel GA, Pfizenmaier K: Sensitivity of leukemic cell lines to cytotoxic alkyl-lysophospholipids in relation to alkyl-cleavage enzyme activities. J Natl Cancer Inst 1987;78:219–222.

6 Fleer EAM, Berkovic D, Eibl H, Unger C: Cellular uptake of hexadecylphosphocholine (HePC). Proc Am Assoc Cancer Res 1990;31:41.

7 Chabot MC, Wykle RL, Modest EJ, Daniel LW: Correlation of ether lipid content of human leukemia cell lines and their susceptibility to 1-0-Ocatdecyl-2-0-methyl-rac-glycero-3-phosphocholine. Cancer Res 1989;49:4441–4445.

8 Vallari DS, Smith ZL, Snyder F: HL60 cells become resistant towards antitumor ether-linked phospholipids following differentiation into a granulocytic form. Biochem Biophys Res Commun 1988;156:1–8.

9 Berkovic D, Fleer EAM, Menzel WA, Unger C: Effect of hexadecylphosphocholine (HePC) on phospholipid metabolism of leukemic Raji cells. Proc Am Assoc Cancer Res 1991.

10 Zheng B, Oiski K, Shoji M, Eibl H, Berdel WE, Hajdu J, Vogler WR, Kuo JF: Inhibition of protein kinase C (sodium plus potassium)-activated adenosine triphosphatase, and sodium pump by synthetic phospholipid analogues. Cancer Res 1990; 50:3025–3031.
11 Herrmann DBJ, Neumann HA: Cytotoxic ether phospholipids. J Biol Chem 1986; 261:7742–7747.
12 Vogler WR, Whigham E, Bennett WD, Olson AC: Effect of alkyl-lysophospholipids on phosphatidylcholine biosynthesis in leukemic cell lines. Exp Hematol 1985;13:629–633.
13 Modolell M, Andreesen R, Pahlke W, Brugger U, Munder PG: Disturbance of phospholipid metabolism during the selective destruction of tumor cells induced by alkyl-lysophospholipids. Cancer Res 1979;K39:4681–4686.
14 Seewald MJ, Olsen RA, Sehgal I, Melder DC, Modest EJ, Powis G: Inhibition of growth factor-dependent inositol phosphate Ca^{2+} signalling by antitumor ether lipid analogues. Cancer Res 1990;50:4458–4463.
15 Oishi K, Raynor RL, Charp PA, Kuo JF: Regulation of protein kinase C by lysophospholipids. J Biol Chem 1988;263:6865–6871.
16 Helfman DM, Barnes KC, Kinkade JM, Vogler WR, Shoji M, Kuo JF: Phospholipid-sensitive Ca^{2+}-dependent protein phosphorylation system in various types of leukemic cells from human patients and in human leukemic cell lines HL60 and K562, and its inhibition by alkyl-lysophospholipid. Cancer Res 1983;43:2955–2961.
17 Kramer IM, Van der Bend RL, Tool ATJ, van Blitterswijk WJ, Roos D, Verhoeven AJ: 1-0-Hexadecyl-2-0-methylglycerol a novel inhibitor of protein kinase C, inhibits the respiratory burst in human neutrophils. J Biol Chem 1989;264:5876–5884.
18 Shoji M, Raynor RL, Berdel WE, Vogler WR, Kuo JF: Effects of thioether phospholipid BM41.440 on protein kinase C and phorbol ester-induced differentiation on human leukemia HL60 and KG-1 cells. Cancer Res 1988;48:6669–6673.
19 Honma Y, Kasukabe T, Okabe-Kado J, Hozumi M, Tusushima S, Nomura H: Antileukemic effect of alkylphospholipids. I. Inhibition of proliferation and induction of differentiation of cultured myeloid leukemia cells by alkylethyleneglycophospholipids. Cancer Chemother Pharmacol 1983;11:73.
20 Honma Y, Kasukabe T, Okabe-Kado J., Hozumi M, Tusushima S, Nomura H: Induction of differentiation of cultured human and mouse myeloid cells by alkyllysophospholipids. Cancer Res 1981;41:3211–3216.
21 McNamara MJC, Schmitt JD, Wykle RL, Daniel LW: 1-0-Hexadecyl-2-acetyl-sn-glycerol stimulates differentiation of HL60 human promyelocytic leukemia cells to macrophage-like cells. Biochem Biophys Res Commun 1984;122:824–830.
22 Hilgard P, Harleman JH, Voegeli R, Maurer HR, Echarti C, Unger C: The antineoplastic activity of hexadecylphosphocholine is associated with tumor cell differentiation. Proc Am Assoc Cancer Res 1989;30:3210.
23 Hochhuth C, Berkovic D, Eibl H, Unger C, Doenecke D: Effects of antineoplastic phospholipids on parameters of cell differentiation in U937 cells. J Cancer Res Clin Oncol 1990;116:459–466.

Dinko Berkovic, Medizinische Universitätsklinik, Abteilung Hämatologie/Onkologie, D-W-3400 Göttingen (FRG)

Eibl H, Hilgard P, Unger C (eds): Alkylphosphocholines: New Drugs in Cancer Therapy. Prog Exp Tumor Res. Basel, Karger, 1992, vol 34, pp 69–76

Hexadecylphosphocholine Amplifies the Effect of Granulocyte Colony-Stimulating Factor on Differentiating Hematopoietic Progenitor Cells

Katalin Vehmeyer,[a] *Torsten Liersch,*[a] *Hansjörg Eibl,*[b] *Clemens Unger*[a]

[a]Department of Internal Medicine, Division of Hematology/Oncology, and
[b]Max-Planck-Institute for Biophysical Chemistry, Göttingen, FRG

Introduction

Membrane-active alkylphosphocholines in principle open new possibilities in the cancer therapy [1–6]. Although these substances are being tested clinically [7, 8], only few data are as yet available concerning biochemical mechanisms mediating the effect of these drugs [9, 10]. On the other hand, it is known that some of the alkylphosphocholines exhibit important biological activities [11, 12]. Thus, it would be reasonable to ask what the functions are of antitumoral effective substances of this group. Do they simply 'overrun' physiological pathways, ignoring their regulating purposes or do they interact biologically with these pathways. Further queries could be: Is the aim of HePC therapy to destroy tumor cells, or should it modify them in the sense of cell differentiation, thereby re-integrating tumor cells in the organism? Could it be that HePC interacts with the cell differentiation?

Most of the modern therapy strategies are clearly oriented to destroy tumor cells (cytotoxic effect) by targeting the DNA or cell membrane or alternatively by disrupting pathways regulating the growth of tumor cells [13, 14]. However, these treatments do not act selectively. They destroy proliferating cells, regardless of their origin. One of the unwelcome problems is the negative effect of cytostatic drugs on the permanently regenerating hematopoiesis [15]. Basically, the intention of this work was to elucidate whether or not hexadecylphosphocholine might have an effect in the tumor therapy in a biologically sensitive way. Experimental and clinical observations with HePC pointed to an appropriate biological system to study this question. Whereas most cytostatic drugs attack hematopoiesis, therapy with alkylphosphocholines tends to result rather in a leukocytosis. Thus, the hematopoiesis has been found to be a valid model for studying the effect of HePC on growing cells.

Hematopoiesis as a Tool of Studying Growth Hormone-Mediated Cell Differentiation

Hematopoiesis is a very sensitive system enabling the replacement of the permanent loss of mature hematopoietic cells. In the growth of hematopoietic cells there is in fact a very fine balance between cell death, cell proliferation and cell differentiation. The development of new hematopoietic cells requires the availability of an intact stem pool, of specific growth hormones and of an appropriate microenvironment. The first step in the growth of hematopoietic cells is the binding of specific growth regulatory substances to their high-affinity receptors. In the regulation of cell growth (proliferation and differentiation) principally four growth mediators take part. These substances are known as colony-stimulating factors (CSF) indicating one of their biological effects in vitro. They support the colony formation of hematopoietic progenitor cells in different stages of maturation. Interleukin-3 (IL-3) or multi-CSF stimulates the growth of primitive progenitor cells (multipotential progenitor cells). The effect of the granulocyte-macrophage colony-stimulating factor (GM-CSF) is more restricted. It mediates the development of early granulocyte-macrophage progenitor cells. The lineage-specific granulocyte colony-stimulating factor (G-CSF) acts preferentially upon neutrophils and their late precursors. Finally, M-CSF (the macrophage colony-stimulating factor) stimulates primarily the differentiation of progenitor cells committed to the macrophage lineage. For the sake of simplicity it could be said that IL-3 interacts mainly with the proliferation of undifferentiated precursors, G-CSF and M-CSF mediate the differentiation of appropriate progenitor cells, and that GM-CSF plays an intermediate role. The above system was used in this study to find out which part of the hematopoiesis might be affected by HePC and how this interaction may occur.

Materials and Methods

Reagents

HePC was provided by ASTA Pharma AG, Frankfurt/Main, FRG.

Preparation of Mononuclear Cells

Fresh samples of bone marrow or peripheral blood were collected in tubes containing 500 U/ml Heparin Novo (Novo Industrie, Mainz, FRG). Samples were diluted 1:2 in Hanks' balanced salt solution (HBSS) and layered over Lymphoprep (Nycomed, Oslo, Norway). After centrifugation at 400 *g* for 40 min, cells of the interphase were collected in HBSS and washed three times. The last washing step was performed at 200 *g* to reduce thrombocyte contamination. Accessory cells were removed according to the method of Law et al. [16]. Briefly, Lymphoprep isolated MNCs were rosetted for 1 h at 4 °C with AET (2-aminoethylisothiouronium bromide hydrobromide, Sigma, Munich, FRG)-treated sheep red blood cells, and then fractioned on Percoll (Biochrom, Berlin, FRG) density gradient. The resulting E-rosette-negative cells were washed twice in HBSS and adjusted to 1×10^7/ml in HBSS with 1% FCS. Monocytes were removed using *L*-phenylalanine methyl ester (PME, Sigma, Munich, FRG). After 40 min of incubation at room temperature the cell suspension was mixed with Percoll to achieve a final Percoll concentration of 25% and layered over 100% Percoll for density gradient separation. The washed cells were then treated with Campath-1 (Serotec, Oxford, UK) antibody to lyse T cells, B cells and monocytes. The antibody was added at a final concentration of 100 μg/ml followed by Percoll separation. Cell counts were determined in a hemacytometer using acridine orange for vital staining.

Agar Cultures

The agar cultures were prepared according to the method of Metcalf [17]. Briefly, $2-5 \times 10^4$ cells, growth factors and 1 ml agar culture-medium consisting of equal volumes of double-strength Dulbecco's modified Eagle's medium (DMEM, Gibco, Karlsruhe, FRG) and 0.6% Bacto-agar (Difco, Detroit, Mich., USA) in distilled water were plated in 35-mm Petri dishes. In all cultures 20% fetal calf serum (FCS, Gibco, Karlsruhe, FRG) was present for the whole incubation period. The serum was pretested for this culture system and was used for all experiments presented in this paper. The recombinant human growth factors GM-CSF (1,200 U/ml), IL-3 (300 U/ml), and G-CSF (130 U/ml) were gifts from Behringwerke (Marburg, FRG). The concentration of growth factors means final concentration. Doses were determined from dose-response curves and were used as a plateau dose of each factor alone or in combination. The plates were incubated at 37 °C under a fully humidified atmosphere and 9% CO_2. Colonies were counted after 10 or 14 days with a dissection microscope (Zeiss, Göttingen, FRG). Cell aggregates containing 50 cells or more were scored as colonies.

Morphological Analysis of the Colonies

Glutaraldehyde-fixed agar cultures were gently floated in tap water onto a microscopic glass slide and covered with a wet Whatman filter paper. The air-dried preparation was stained with 0.5% Solvent Blue 38 (Sigma, Munich, FRG), dissolved in 70% urea-saturated ethanol for 2 h, and after washing in water subsequently counterstained with hematoxylin for 15 min.

Patients

Three patients with Philadelphia chromosome positive chronic phase CML and 31 patients without known hematological disorders, admitted for orthopedic surgery, were used as donors. CML patients were without treatment for at least 8 weeks before progenitor cells were taken from the peripheral blood. The 31 donors without known hematological disorders were elderly patients (60–80 years) suffering from severe coxarthrosis. All had a history of long medication with analgetics and/or non-steroidal anti-inflammatory drugs. Bone marrow samples were obtained during the surgery.

Results

Colony formation was investigated in the presence of the hematopoietic growth factors G-CSF, GM-CSF and IL-3 either as a single agent or in combination. Recombinant human G-CSF, GM-CSF and IL-3 were incorporated at plateau concentrations previously determined to have maximal colony-stimulating activity.

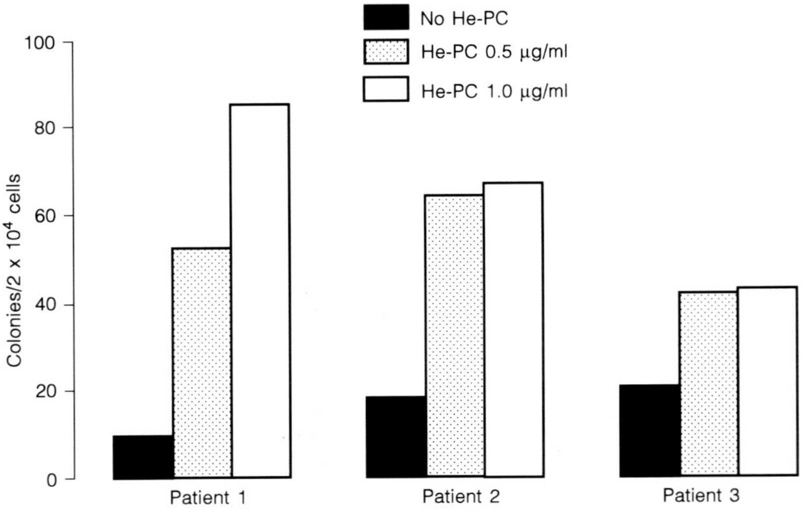

Fig. 1. Effect of HePC on GM-progenitor cells in the presence of G-CSF. Patients suffering from chronic myeloid leukemia.

Incubations were carried out by omission of HePC and in the presence of increasing concentrations of the compound ranging from 0.5 to 8 μg/ml. Without G-CSF, GM-CSF and IL-3, HePC failed to induce colony formation. Thus, this substance has no mitogenic activity by itself.

Hematopoietic progenitor cells of 3 patients suffering from CML and of 31 patients without known hematopoietic disorders were investigated. Summarizing the data, CSF-dependent colony formation of progenitor cells from CML patients were stimulated in the presence of 0.5 and 1 μg HePC/ml (fig. 1). An impressive effect was observed in the case of G-CSF where the stimulating activity was up to 8 times greater than that of the controls (without HePC). GM-CSF- and IL-3-induced colony growth was increased only up to 2-fold, whereas marginal effects were noted in cultures supplemented with a combination of the different CSFs [data not shown]. Higher concentrations of HePC, 4 and 8 μg/ml, strongly suppressed the growth of G-CSF- and GM-CSF-induced colony formation. However, even at 8 μg HePC/ml [data not shown], no suppressive effect below the control value could be detected when IL-3 was used as a single growth factor.

Also in the group of patients without known hematological disorders HePC showed a stimulatory effect. Sixteen out of 31 patients acted as HePC responders, 15 patients were nonresponders. This indicates that

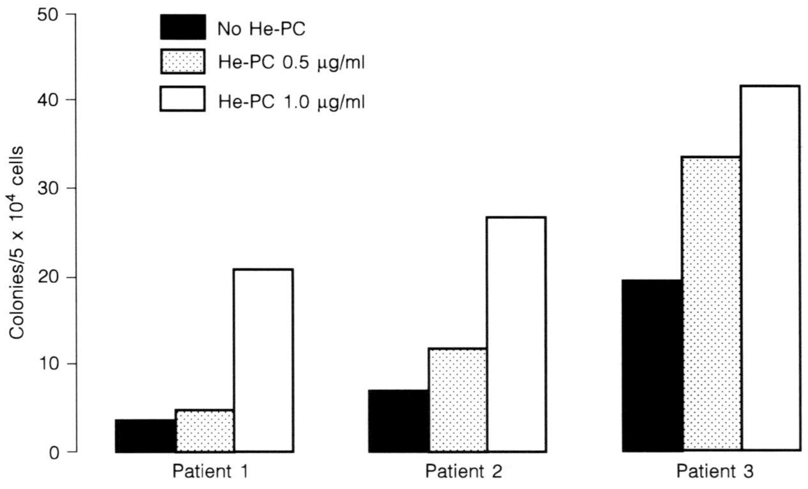

Fig. 2. Effect of HePC on GM-progenitor cells in the presence of G-CSF. Patients without known hematological disorders.

the biological state of progenitor cells determines both the CSF and the HePC effect. Figure 2 shows the results obtained from three representative responder patients without hematological disorders. Again, HePC stimulated the G-CSF-induced colony formation up to 8-fold. Borderline or no effect was obtained in the case of GM-CSF or IL-3 [data not shown]. As could be observed in the case of CML patients, higher amounts of HePC had strong suppressive effects on colony formation when cultures were grown in the presence of G-CSF and GM-CSF, whereas no or only a slight suppression was noted in the presence of IL-3.

Morphological analysis of the colonies revealed that neither the stimulating effect of HePC (0.5–2 μg/ml) nor the inhibition by HePC at a higher concentration (8 μg/ml) on colony growth of hematopoietic progenitors could be attributed to a shift in the growth of specific colonies. The cell composition of the individual colonies and the distribution of the different colony types were found to be identical with those of the control cultures [data not shown]. Thus, HePC at low doses appears to act as a co-stimulator especially with G-CSF. At higher doses it suppresses the G-CSF and GM-CSF-dependent development of colonies but it hardly interacts with the IL-3-mediated colony growth.

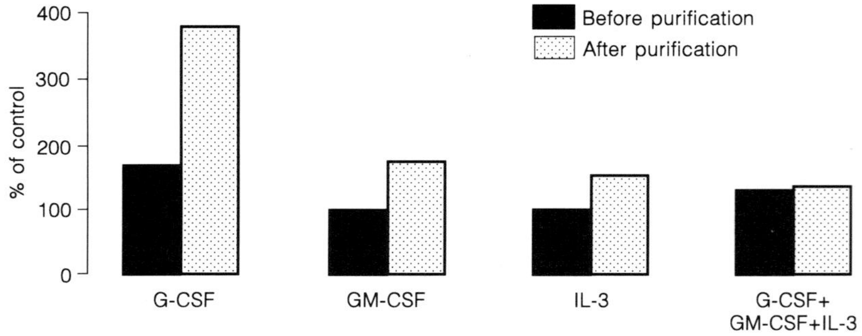

Fig. 3. GM-CFU-c with and without accessory cells in the presence of 0.5 μg HePC/ml in a case of CML.

Studies on progenitor cells after depletion of accessory cells were performed in 2 cases. Isolated progenitor cells of peripheral blood of a patient suffering from myeloproliferative disorder and a patient without known hematological disorder were investigated. Figure 3 shows the CSF-dependent progenitor cell growth of the patient suffering from CML before and after the removal of accessory cells. HePC was very effective as a co-stimulator in cultures both with and without accessory cells (fig. 3). The stimulatory potential of HePC was actually higher in the cultures depleted of accessory cells (fig. 3).

In the case of the patient without hematological disorders there was a similar effect detectable [data not shown]. Again, the stimulatory potential of HePC was higher in the cultures without accessory cells. These results strongly indicate that the co-stimulatory effect of HePC in the CSF-dependent colony growth of hematopoietic progenitors was unlikely to depend on accessory cell activity.

Discussion

The in vitro assay system in this study allowed us to investigate not only the hematopoiesis but also the regulated cell proliferation/cell differentiation.

In this experimental system a selective co-operative effect of low-dose HePC on the growth of hematopoietic cells was observed. HePC (0.5–1 μg/ml) acted only in the presence of specific growth hormones as G-CSF, GM-CSF and/or IL-3. Thus, HePC operates clearly as a co-stimulator. Furthermore, it has been found that the action of HePC corresponds above all to the effect of G-CSF, representing a typical differentiation specific growth stimulus. In the presence of G-CSF, HePC stimulated a colony growth of granulocytes/macrophages up to 8-fold. The cooperation with GM-CSF resulted at the most in a 2-fold enhanced colony growth. Interestingly, HePC showed hardly an effect in cooperation with IL-3 which stimulates above all the renewal of pluripotent stem cells.

Higher concentrations of HePC (4–8 μg/ml) were inhibitory in this culture system. However, only a minimum decrease of the colony growth by HePC could be observed in cultures supplemented with IL-3, showing again a selective effect of HePC.

The mechanism both of the stimulatory effect of HePC at lower concentrations and of its inhibitory property at higher concentrations is unknown. Accessory cells seem not to be involved in these processes since the removal of accessory cells did not eliminate the effect of HePC.

As a further possibility, He-PC-enhanced expression of high-affinity receptors for CSFs should be discussed. HePC has been found to enhance the expression of IL-2 receptors on T cells [18]. Nevertheless, this possibility in the case of hematopoietic progenitor cells remains to be elucitated.

The participation of HePC in signal transduction pathways as a third possible mechanism is also speculative for the time being. There are some observations describing the interaction of HePC with the protein kinase C (PKC) [9, 10]. However, detailed investigations which would reduce all these observations to a common denominator are still missing.

Although the mechanism of the effect of HePC is unknown, the fact that HePC interacts predominantly with a growth hormone which chiefly regulates the cell differentiation is surprising. Indeed, data presented in this communication could point to a completely different mode of application of alkylphosphocholines in tumor therapy.

References

1 Eibl H, Unger C: Hexadecylphosphocholine, a new and selective antitumor drug. Cancer Treat Rev 1990;17:233–242.

2 Unger C: Alkylphosphocholine und Analoga: Entwicklung einer neuen Substanzgruppe mit antineoplastischer Wirkung; Stuttgart, Thieme, 1989.

3 Unger C, Damenz W, Fleer EAM, Kim DJ, Breiser A, Hilgard P, Engel J, Nagel G, Eibl H: Hexadecylphosphocholine, a new ether lipid analogue. Studies on the antineoplastic activity in vitro and in vivo. Acta Oncol 1989;28:213–217.

4 Hilgard P, Stekar J, Voegeli R, Engel J, Schumacher W, Eibl H, Unger C, Berger MR: Characterization of the antitumor activity of hexadecylphosphocholine (D 18506). Eur J Cancer Clin Oncol 1988;24:1457–1461.

5 Muschiol C, Berger MR, Schuler B, Scherf HR, Garzon FT, Zeller WJ, Unger C, Eibl H, Schmähl D: Alkylphosphocholines: Toxicity and anticancer properties. Lipids 1987;22:930–934.

6 Schick HD, Berdel WE, Fromm M, Fink U, Jehn U, Ulm K, Reichert A, Eibl H, Unger C, Rastetter J: Cytotoxic effects of ether lipids and derivatives in human nonneoplastic marrow cells and leukemic cells in vitro. Lipids 1987;22:904–910.

7 Unger C, Eibl H, Nagel GA, von Heyden HW, Breiser A, Engel J, Steker J, Peukert M, Hilgard P, Berger M: Hexadecylphosphocholine in the topical treatment of skin metastases: A phase I trial. Onkologie 1989;37:219–223.

8 Unger C, Eibl H, von Heyden HW, Peukert M, Sindermann H, Nagel GA: Phase I study with daily hexadecylphosphocholine in patients with malignant disease. J Cancer Res Clin Oncol 1990;116 (suppl):993.

9 Grunicke H, Hofmann J, Oberhuber H, Überall F, Zaknun J, Voegeli R, Hilgard P: Hexadecylphosphocholine inhibits protein kinase-C and depresses the inositol phosphate response in NJH 3T3 fibroblasts. J Cancer Res Clin Oncol 1990;116 (suppl):889.

10 Kuo JF, Zheng B, Shoji M, Vogler WR, Eibl H: Inhibition of protein kinase C, Na, K-ATPase and HL60 cell differentiation by alkylphosphocholine and alkylammonium bromide derivatives. J Cancer Res Clin Oncol 1990;116 (suppl):889.

11 Beneviste J, Hensen PM, Cochrane CG: Leukocyte-dependent histamine release from rabbit platelets. The role of Ig E, basophils and a platelet activating factor. J Exp Med 1972;136:1356–1377.

12 Lee T-C, Snyder F: Function, metabolism and regulation of platelet activating factor and related ether lipids; in Kuo (ed): Phospholipids and Cellular Regulation. Boca Raton, CRC Press, 1985, vol 2, pp 1–39.

13 Calabresi P, Parks RE: Chemotherapy of neoplastic diseases; in Goodman LS, Gilman A (eds): The Pharmacological Basis of Therapeutics. New York, MacMillan, 1980.

14 Tritton TR, Hickman JA: How to kill cancer cells: Membranes and cell signalling as targets in cancer chemotherapy. Cancer Cells 1990;2:95–105.

15 Pizzo PA: Granulocytopenia and cancer therapy. Cancer 1984;54:2649–2661.

16 Law P, Chau W-K, Dooley DC: Purification of peripheral blood hematopoietic cells using Campath-1M monoclonal antibody. Exp Hematol 1990;18:289–293.

17 Metcalf D: The Hemopoietic Colony Stimulating Factors. Amsterdam, Elsevier, 1984, pp 110–112.

18 Vehmeyer K, Scheurich P, Eibl H, Unger C: Hexadecylphosphocholine-mediated enhancement of T cell responses to interleukin 2. Cell Immunol 1991; in press.

Katalin Vehmeyer, Department of Internal Medicine, Division of Hematology/Oncology, Robert-Koch-Strasse 40, D-W-3400 Göttingen (FRG)

Eibl H, Hilgard P, Unger C (eds): Alkylphosphocholines: New Drugs in Cancer Therapy. Prog Exp Tumor Res. Basel, Karger, 1992, vol 34, pp 77–89

Early Effects of Hexadecylphosphocholine on Gene Expression in Leukemia Cell Lines

C. Hochhuth[a], *D. Doenecke*[a], *C. Unger*[a], *H. Eibl*[b]

[a]Medizinische Universitätsklinik, Abteilung Hämatologie/Onkologie, and [b]Max-Planck-Institut für biophysikalische Chemie, Göttingen, FRG

Introduction

The alkylphosphocholine hexadecylphosphocholine (HePC) shows a variety of biological properties. HePC inhibits the proliferation of several tumor cell lines in a dose-dependent manner [1–3]. Besides this antiproliferative activity, HePC can induce differentiation of leukemic cells into mature myeloid cells [4] and some data indicate that induction of differentiation may contribute to the in vivo antitumoral activity of HePC in chemically induced rat mammary carcinomas [2]. The process of differentiation is associated with numerous changes in the pattern of cellular proteins, for example cell-surface antigens as well as structural and regulatory polypeptides. This alteration of the protein pattern in differentiated cells often depends on altered gene expression.

Human promyelocytic and promonocytic leukemia cell lines present excellent systems for the investigation of such changes in gene expression during the process of differentiation. The human promyelocytic cell line HL-60 can be induced to differentiate into cells of the granulocytic lineage by dimethylsulfoxide (DMSO) or retinoic acid [5, 6], whereas phorbol ester and vitamin D_3 analogues induce differentiation into cells with a monocytic phenotype [6]. Both types of induction are associated with growth inhibition. The proliferation of HL-60 cells can also be inhibited by HePC [3]. Besides its antiproliferative activity, HePC induces differentiation of HL-60 cells towards mature cells of the granulocyte lineage [unpubl. data]. The human promonocytic cell line U937 can be induced to differentiate to monocyte/macrophage-like cells by phorbol ester or vitamin D_3 analogues [7]. HePC also has the capacity to trigger monocyte differentiation in U937 cells [8]. During the course of differentiation the pattern of surface antigen expression changes towards a granulocyte- (HL-60) or monocyte-like (U937) phenotype [8]. This includes for example the acquisition of

CD14, CD11 and HLA-DR antigen, whereas transferrin receptors and CD4 antigen as typical markers for early stages of the myeloic lineage are reduced or lost [9]. At least 48 h of incubation in the presence of inducing compounds are necessary to investigate these parameters of differentiation. The interpretation of such rather late events is sometimes difficult, because it cannot be excluded that the observed effects are epiphenomenons and that they are not primarily caused by the tested substance. Therefore, we investigated *early* effects of HePC on human leukemia cell lines in order to elucidate the mechanism of HePC activity. The interaction and incorporation of HePC within the cell membrane results in various biological effects on several cellular processes and finally in an inhibition of DNA replication. An arrest of DNA synthesis is frequently accompanied by changes of the pattern of chromosomal proteins [10, 11]. Nondividing, terminally differentiated cells, such as hepatocytes [12–14] or nerve cells [15, 16], show an increased level of the H1 histone subtype $H1^0$. Unlike other histones, the synthesis of this subtype does not depend on DNA replication [17]. It has been demonstrated in a variety of systems that several inhibitors of cell growth induce de novo synthesis of this chromosomal protein [17–21]. For example, treatment of HeLa cells with sodium butyrate results in an accumulation of $H1^0$ mRNA after 3–6 h of incubation followed by an increase of $H1^0$ protein after 24 h [22]. Therefore, the expression of the $H1^0$ gene presents one parameter to study the early effects of HePC on leukemia cell lines. We described the kinetics of $H1^0$ mRNA accumulation in the human promonocytic cell line U937 after addition of the alkylphosphocholine HePC. However, the $H1^0$ gene is not transcribed in all cell types. While unstimulated U937 cells contain only small amounts of $H1^0$ transcripts, no $H1^0$ mRNA can be detected in untreated HL-60 cells. Sodium butyrate, serum starvation or density inhibition induce enhanced $H1^0$ gene expression in several cell systems [20], but none of these methods results in measurable $H1^0$ gene transcription in HL-60 cells. The reason for the complete inactivity of the $H1^0$ gene in this cell type is yet not defined. It might be related to the lineage commitment of HL-60 cells to granulocytes. Lymphocytes and granulocytes do not express this H1 subtype [11].

We further investigated the effect of HePC on the expression of the c-myc oncogene, which is down-regulated in differentiating leukemia cells upon phorbol ester treatment [23]. The c-myc oncogene codes for a nuclear protein and its expression is correlated with cell proliferation [24]. The expression of core histone genes and H1 genes of the main type are restricted to the S-phase of the cell cycle and in this way are also correlated with cell proliferation [25, 26]. Therefore, the inhibition of proliferation is accompanied by a decline of histone mRNA concentration, but this decreased gene expression appears later than the down-regulation of the c-myc oncogene. Induction of differentiation in both cell lines is associated with an early decrease of c-myc mRNA concentration [23].

The accumulation of $H1^0$ mRNA in U937 cells after addition of HePC is one example for the induction of enhanced gene expression, whereas the decrease of c-myc transcript concentration under the same conditions is an example for down-regulation of another gene. Both effects are early events during the process of differentiation and inhibition of cell proliferation induced by HePC. In comparison to HePC, we analyzed the effects of hexadecylphospho-(N,N,N-trimethyl)-hexanolamine (He-PC_6) on gene expression in U937 and HL-60 cells. This substance shows very similar physical properties with respect to oil/water partition coefficients (log p values), critical micelle concentration and induction of hemolysis in human erythrocytes [27], but it lacks antitumoral activity and has no effect on cell proliferation.

Cells and Methods

Cell Culture

Human promyelocytic leukemia HL-60 cells and promonocytic U937 cells were continuously cultured in RPMI-1640 medium, supplemented with 10% fetal calf serum, penicillin/streptomycin (100 U/ml and 100 g/ml, respectively) and glutamine (2 m*M*) at 37 °C, 5% CO_2. Cells were seeded at a density of 2×10^5 cells/ml and incubated in the presence or absence of test substances for different periods of time.

RNA analysis

Cells were washed twice in ice-cold phosphate-buffered saline (0.14 *M* NaCl, 2.5 m*M* KCl, 8.1 m*M* HPO_4, 1.5 m*M* KH_2PO_4, pH 7.4) and total cellular RNA was prepared by the guanidinium isothiocyanate/cesium chloride method [28]. Cell homogenates in lysis buffer (4 *M* guanidinium isothiocyanate, 0.5% (w/v) N-lauroylsarcosine, 25 m*M* sodium citrate, 0.1 *M* mercaptoethanol) were centrifuged in a Beckman SW 60 rotor for 16 h at 180,000 *g* (20 °C). The RNA sediment was dissolved in sodium dodecyl sulfate buffer (10 m*M* Tris/HCl, pH 7.2, 5 m*M* EDTA, 1% (w/v) sodium dodecyl sulfate), precipitated with ethanol, washed several times with 80% ethanol, dried and dissolved in water. Samples of 10 μg RNA were denatured in glyoxal (5%, v/v) and dimethylsulfoxide (50%, v/v) for 1 h at 50 °C [29]. Denatured RNA was electrophoretically separated on 1% agarose gels, transferred onto nylon membranes (Amersham Hybond N) and hybridized with labeled DNA probes as previously described [22]. Endogenous ribosomal RNA could be used as a marker, since human $H1^0$ mRNA, being about 2.3 kb in length [30], has a size intermediate between 28S and 18S RNA, whereas main-group H1 measures around 0.7 kb [31]. DNA probes were derived from the human $H1^0$ gene [30], from two human main-type H1 genes [31] and from human core histone genes [31]. The DNA probe for the human c-myc oncogene was 1.5 kb EcoRI-HindIII fragment coding for parts of the first intron and the second exon (Amprobe, Amersham). After labelling the DNA fragments by random priming (Multiple DNA labelling system, Amersham), the probes were used for hybridization.

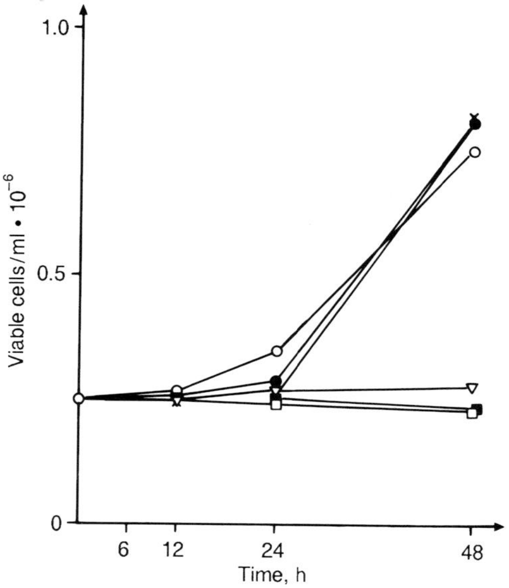

Fig. 1. Effects of various phospholipids, phospholipid analogues and phorbol ester on growth kinetics of U937 cells: 1-O-octadecyl-2-O-methyl-rac-glycero-3-phosphocholine (ET-18-OCH_3) (□) 5 μg/ml, hexadecylphosphocholine (HePC) (▽) 10 μg/ml, hexadecylphospho-(N,N,N-trimethyl)-hexanolamine ($HePC_6$) (*) 10 μg/ml, 1-O-palmitoyl-sn-glycero-3-phosphocholine (ES-16-OH) (●) 10 μg/ml, and 12-O-tetradecanoylphorbol 13-acetate (TPA) (■) 32 n*M* were added to U937 cells at an initial density of 2×10^5 cells/ml. The *ordinate* indicates the number of viable cells/ml and the *abcissa* refers to the time interval after additon of the respective substance. O = Untreated cells.

Results

Effects of Phospholipids, Alkylphosphocholines and Phorbol Ester on U937 Cell Proliferation

U937 cells are derived from a human promonocytic leukemia [32]. They can be induced to differentiate towards monocyte-like cells upon addition of a number of different compounds like phorbol ester, retinoic acid, interferon-gamma, colony-stimulating factors and tumor necrosis factor-alpha. Typically, these U937 cells cease to proliferate and, being treated with some substances, become attached to the culture-vessel walls. We have treated U937 cells with phorbol ester TPA, the alkylphosphocholine HePC, hexadecylphospho-(N,N,N-trimethyl)-hexanolamine (He-PC_6), the acyllysophospholipid ES-16-OH and the alkyllysophospholipid ET-18-OCH_3. Figure 1 [taken from ref. 8] shows the influence of these compounds on U937 cell proliferation. During an initial phase of 24 h cell division does not occur. Subsequently, control cells start to divide resulting in fourfold cell number at 48 h. Cells treated with HePC, ET-18-OCH_3 and TPA, however, do not proliferate, whereas those treated with $HePC_6$ and ES-16-OH show growth characteristics similar to control cells.

Effects of HePC, $HePC_6$, ES-16-OH, ET-18-OCH_3 and Phorbol Ester on HL – 60 Cell Proliferation

The human cell line HL-60 was established from a patient with acute promyelocytic leukemia [33]. Differentiation of HL-60 cells towards cells with a granulocyte phenotype can be induced by dimethylsulfoxide (DMSO) or retinoic acid, whereas treatment with phorbol ester results in differentiation to monocyte/macrophage-like cells [6]. The process of differentiation is accompanied by an inhibition of cell proliferation. We have treated HL-60 cells with TPA, HePC, $HePC_6$, ES-16-OH and ET-18-OCH_3. Figure 2 demonstrates the effects of these compounds on HL-60 cell proliferation. The growth kinetics of HL-60 cells in the presence of the tested substances are similar to those of U937 cells, which are shown in figure 1. But the antiproliferative activity of HePC and ET-18-OCH_3 in HL-60 cells is more pronounced and precedes the cytotoxic effect observed in U937 cells.

Effects of HePC and $HePC_6$ on H1 Gene Expression in U937 and HL-60 Cells

The synthesis of the majority of histones is tightly coupled to DNA replication. However, the expression of a few histone subtypes does not depend on DNA replication. These few subtypes are normally restricted to terminally differentiated cells with a low proliferation rate. We analyzed steady state levels of different H1 histone mRNA species in U937 and HL-60 cells after addition of HePC, $HePC_6$, alkyllysophospholipid ET-18-OCH_3 and acyllysophospholipid ES-16-OH. The H1 subtype $H1^{II}$ represents a replication-dependent H1 gene. Therefore, the stop of cell proliferation is associated with a decrease of $H1^{II}$ transcript concentration after about 24 h of addition of compounds with antiproliferative activity. The level of $H1^0$ mRNA is very low in proliferating U937 cells, whereas no $H1^0$ transcripts can be detected in dividing HL-60 cells. Incubation of U937 cells with the alkylphosphocholine HePC results in a drastic accumulation of $H1^0$ mRNA which can already be measured 1–3 h after addition of HePC to the culture medium. Figure 3 [taken from ref. 8] shows the increase of $H1^0$ transcript concentration induced by HePC, whereas the amounts of the S-phase regulated $H1^{II}$ subtype and core histone mRNA decrease after 24 h of incubation. The significant rise of $H1^0$ mRNA concentration is also observed in U937 cells upon addition of the

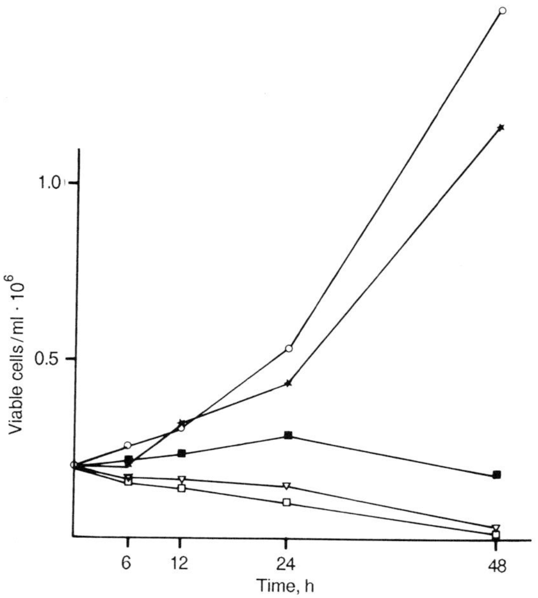

Fig. 2. Effects of phospholipid analogues and phorbol ester on growth kinetics of HL-60 cells: 1-O-octadecyl-2-O-methyl-rac-glycero-3-phosphocholine (ET-18-OCH_3) (□) 5 μg/ml, hexadecylphosphocholine (HePC) (▽) 10 μg/ml, hexadecylphospho-(N,N,N-trimethyl)-hexanolamine ($HePC_6$) (*) 10 μg/ml, and 12-O-tetradecanoylphorbol 13-acetate (TPA) (■) 32 n*M* were added to HL-60 cells at an initial density of 2×10^5 cells/ml. The *ordinate* indicates the number of viable cells/ml and the *abscissa* refers to the time interval after addition of the respective substance. O = Untreated cells.

alkyllysophospholipid ET-18-OCH_3 [8]. Like HePC, ET-18-OCH_3 treatment leads to an early rise of $H1^0$ mRNA concentration with a maximum after 12–24 h of incubation. Prolonged cultivation results in a decrease of $H1^0$ transcript concentration.

The amount of S-phase-dependent H1 and core histone transcripts declines about 24 h after addition of HePC or ET-18-OCH_3. This correlates well with the cessation of cell proliferation. Figure 3 demonstrates that the effect of HePC treatment is not reversible after 24 h of incubation: removal of inducer after this time and further cultivation of the cells in control medium does not abolish the observed HePC effects.

Similar effects on histone gene expression are observed in U937 cells after incubation with ET-18-OCH_3. This substance also shows antiproliferative activity in U937 cells (fig. 1). The accumulation of $H1^0$ mRNA precedes the decrease of S-phase dependent H1 transcripts. In contrast to

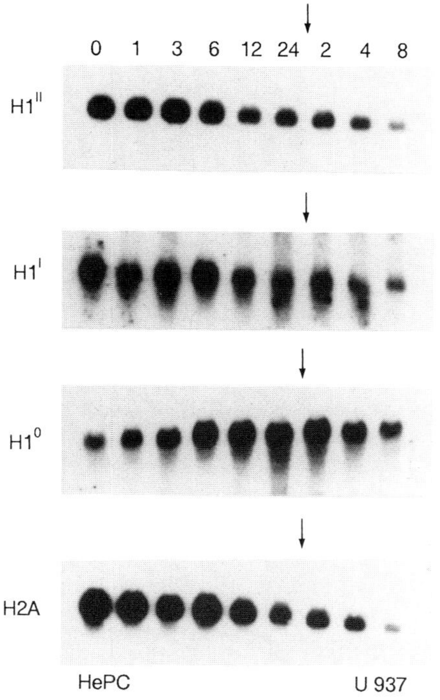

Fig. 3. Steady state levels of mRNA coding for histone H1^0 and cell-cycle-dependent histone species. RNA was extracted at different times (*top line*, h) after addition of hexadecylphosphocholine (HePC, 25 μM). Identical amounts of RNA (10 μg) were separated electrophoretically, blotted and hybridized with ^{32}P-labeled probes, derived from genes coding for histone H1^0, two different H1 histones and an H2A core histone gene. The *vertical arrow* indicates removal of HePC, washing and resuspension of the cells in HePC-free medium.

ET-18-OCH_3 and HePC, $HePC_6$ and the acyllysophospholipid ES-16-OH do not possess such an antiproliferative capacity in U937 and HL-60 cells.

$HePC_6$ has a very similar structure to HePC and comparable physicochemical properties. Therefore, Northern blot analyses were done to investigate whether this compound can induce enhanced H1^0 gene expression like HePC. ES-16-OH was tested as a second control substance, which does not possess antiproliferative or differentiating activity. The analysis of steady state levels of different H1 subtype transcripts revealed that the concentration of H1 mRNA from main-type H1 genes is only slightly influenced by $HePC_6$ (fig. 4) and ES-16-OH. The concentration of H1^0 mRNA also shows only marginal variations during a 48-hour incuba-

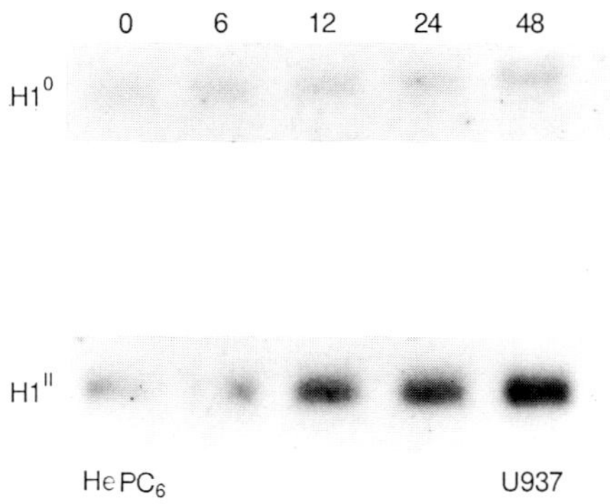

Fig. 4. Levels of mRNA for H1^{0} and H1^{II} in U937 cells, exposed for different periods of time (*top line*, h) to *HePC$_6$* (25 *μM*). Identical amounts of RNA were electrophoresed, blotted and hybridized with the respective probes.

tion time in the presence of HePC$_6$ or ES-16-OH. These data indicate that the antiproliferative activity and the induction of differentiation of the tested compounds correlate with their effects on histone gene expression. Equimolar concentrations of HePC$_6$ neither show an inhibition of cell proliferation nor a significant influence on histone gene expression.

HePC also inhibits the proliferation of HL-60 cells and can induce differentiation into cells of granulocyte phenotype. HePC$_6$ does not possess such antiproliferative and differentiating activity. Northern blot analyses of RNA from HePC$_6$ treated HL-60 cells revealed that the concentration of S-phase dependent H1 histone gene transcripts decreases during HePC incubation, where HePC$_6$ has only marginal and transient effects on H1 gene expression [data not shown]. The expression of the H1^{0} gene in HL-60 cells can neither be induced by HePC, by ET-18-OCH$_3$, nor by sodium butyrate. Until now, it is not defined why the H1^{0} gene is completely inactive in HL-60 cells and can even not be induced by sodium butyrate, a well-known activator of H1^{0} gene expression [18]. Mature granulocytes do not express the H1^{0} gene, and it might be that this histone gene is also totally inactive in granulocyte progenitor cells.

Effects of HePC and HePC$_6$ on c-myc Oncogene Expression in U937 and HL-60 Cells

Besides the influence of HePC on histone gene expression, we investigated the effect of HePC on the expression of the c-myc oncogene in U937 and HL-60 cells. Both cell lines contain rather high amounts of c-myc mRNA. The induction of differentiation in U937 and HL-60 cells is

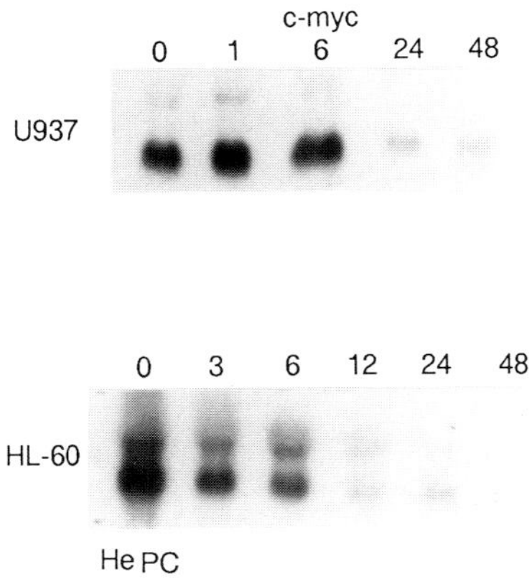

Fig. 5. Effect of HePC (25 μM) on levels of mRNA coding for the c-myc oncogene. Each lane represents identical amounts of RNA, extracted from U937 cells (upper part) and from HL-60 cells (lower part) after the periods of HePC treatment indicated (h) above each series.

accompanied by an early decrease of c-myc transcript concentration. In parallel, we studied steady state levels of c-myc mRNA in both cell lines during incubation with HePC and $HePC_6$. The Northern blot analysis of figure 5 shows that the concentration of c-myc transcripts in U937 cells is transiently increased after 3 h of incubation and decreased afterwards. In HL-60 cells, the basal expression of the c-myc oncogene is higher than in U937 cells. This high expression is not enhanced, but declines early after addition of HePC to the culture medium. In contrast, the amount of c-myc mRNA varies only slightly during the whole time of incubation in the presence of $HePC_6$.

These data show that treatment of the leukemia cell lines U937 and HL-60 with HePC is accompanied by an altered gene expression. These modulations of steady state levels of transcripts of histone genes and of the c-myc oncogene are early events, which can already be measured within the first hours of incubation with HePC. Incubation of both cell lines in the presence of $HePC_6$, which has neither an antiproliferative nor a differentiation-inducing effect, is not associated with such significant changes in gene expression, but results in transient modulations of transcript concentration.

Discussion

The two human leukemia cell lines U937 and HL-60 can be induced to differentiate towards granulocyte (HL-60) or monocytes (HL-60, U937) upon addition of several substances. The alkylphosphocholine HePC also shows antiproliferative activity and can induce differentiation in both cell lines. In contrast, hexadecylphospho-(N,N,N-trimethyl)-hexanolamine ($HePC_6$), which differs from HePC by an enlarged distance between the phosphor and nitrogen atoms, does not lead to inhibition of proliferation and induction of differentiation in U937 and HL-60 cells. Changes in surface antigen and enzyme expression, which are specific for the process of differentiation, appear after several days of incubation. Phorbol ester induced differentiation of leukemia cells is accompanied by an early decrease of c-myc mRNA concentration [23]. We investigated early effects of HePC and $HePC_6$ on the expression of c-myc oncogene and several H1 genes. Northern blot analyses of RNA from HePC-treated U937 and HL-60 cells revealed an altered expression of different histone H1 subtype genes. The concentration of transcripts from main-type H1 genes declines in correlation with the inhibition of cell proliferation, whereas $H1^0$ mRNA accumulates in U937 cells after a short time of incubation. The $H1^0$ gene is neither expressed in untreated nor in HePC-stimulated HL-60 cells. The expression of the c-myc oncogene is down-regulated upon He-PC addition in both cell lines. In contrast, incubation of both cell types in the presence of He-PC_6 does not significantly influence the expression of H1 genes and c-myc oncogene. Phorbol ester stimulation of U937 and HL-60 cells is accompanied by similar effects concerning c-myc and H1 gene expression. One approach to elucidate the mode of HePC activity is the investigation of phospholipid metabolism and membrane-associated signal transduction systems.

A different approach is done here by the study of gene expression of some genes under the influence of HePC. The regulation of gene expression is a rather complex and just partly resolved network of interactions between DNA and protein factors. Some regulatory elements, which are located upstream or downstream of specific genes, have been characterized as yet. These regulatory elements are DNA sequences, which can interact with one or more protein factors and thereby influence the expression of the neighboring gene. Some of these regulatory protein factors have been isolated and characterized [34, 35]. The activity of these factors themselves has to be regulated. In some cases, de novo protein synthesis is necessary to provide such a transcription factor [36]. Activation of a preexisting, but inactive factor is another way to modulate gene expression under certain circumstances [37]. For example, activation of inactive cytosolic transcription factor NF-kappaB occurs by phosphorylation of an inhibitory protein, which builds a complex together with NF-kappaB. This phosphorylation results in a dissociation of the inhibitory protein and the release of active NF-kappaB followed by the translocation of the now active transcription factor into the nucleus [38, 39].

Both positive and negative regulatory elements have been identified for human and murine c-myc genes [40–42] as well as for human and murine $H1^0$ genes [43]. Some of the protein factors which interact with these elements are characterized. In order to elucidate the mechanism of HePC activity, we will investigate those regulatory factors which are activated or become inactive during HePC incubation. The next step will be to study the signal transduction by which these factors themselves are regulated. Thereby, we investigate HePC-induced gene expression in the nucleus, which is regulated by transcription factors and these become, at least in some cases, activated in the cytoplasm. Such data together with results concerning integration of HePC into the cell membrane and effects on phospholipid metabolism, will improve the understanding of biological properties of HePC.

References

1 Muschiol C, Berger MR, Scherf HE, Garzon FT, Zeller WJ, Unger C, Eibl H, Schmähl D: Alkylphosphocholines: Toxicity and anticancer properties. Lipids 1987;22:930–934.

2 Hilgard P, Stekar J, Voegeli H, Engel J, Schumacher W, Eibl H, Unger C, Berger MR: Characterization of the antitumour activity of hexadecylphosphocholine (D18506). Eur J Cancer Res Clin Oncol 1988;28:1457–1461.

3 Unger C, Damenz W, Fleer EAM, Kim DJ, Breiser A, Hilgard P, Engel J, Nagel G, Eibl H: Hexadecylphosphocholine, a new etherlipid analog. Studies on the antineoplastic activity in vitro and in vivo. Acta Oncol 1989;28:213–217.

4 Hilgard P, Harlemann JH, Voegeli R, Maurer HR, Echarti C, Unger C: The antineoplastic activity of hexadecylphosphocholine (HPC) is associated with tumor cell differentiation. Proc Am Assoc Cancer Res 1989;30:2310.

5 Collins SJ, Ruscetti F, Gallagher RE, Gallo RC: Terminal differentiation of human promyelocytic leukemia cells induced by dimethyl sulfoxide and other polar compounds. Proc Natl Acad Sci USA 1978;75:2458–2462.

6 Koeffler PH: Induction of differentiation of human acute myelogenous leukemia cells: Therapeutic implications. Blood 1983;62:709–721.

7 Radzun HJ, Parwaresch MR, Sundström C, Nilsson K, Eissner M: Monocytic origin of the human hematopoetic cell line U937 and its convertibility to macrophages evidenced by isoenzyme mapping. Int J Cancer 1983;31:181–186.

8 Hochhuth C, Berkovic D, Eibl H, Unger C, Doenecke D: Effects of antineoplastic phospholipids on parameters of cell differentiation in U937 cells. J Cancer Res Clin Oncol 1990;116:459–466.

9 Faltyneck CR, Finch LR, Miller P, Overton WR: Treatment with recombinant IFN-gamma decreases cell surface CD4 levels on peripheral blood monocytes and on myelomonocyte cell lines. J Immunol 1989;142:500–508.

10 D'Anna JA, Gurley LR, Tobey RA: Synthesis and modulations in the chromatin contents of histones $H1^0$ and H1 during G_1 and S phases in Chinese hamster cells. Biochemistry 1982;21:3991–4001.

11 D'Incalci M, Allavena P, Wu RS, Bonner WM: H1 variant synthesis in proliferating and quiescent human cells. Eur J Biochem 1986;154:273–279.

12 Panyim S, Chalkley RG: A new histone found only in mammalian tissue with little cell division. Biochem Biophys Res Commun 1969;37:1042–1049.

13 Gjerset R, Gorka C, Hasthrope S, Lawrence JJ, Eisen H: Developmental and hormonal regulation of protein $H1^0$ in rodents. Proc Natl Acad Sci USA 1982;2333–2337.

14 Gabrielli F, Aden D, Carrel SC, von Bahr C, Rane A, Angeletti CA, Hancock R: Histone complements of human tissues, carcinomas and carcinoma derived cell lines. Mol Cell Biochem 1985;65:57–66.

15 Pina B, Martinez P, Simon L, Suau P: Differential kinetics of histone $H1^0$ accumulation in neuronal and glial cells from rat cerebral cortex during postnatal development. Biochem Biophys Res Commun 1984;123:697–702.

16 Pina B, Suau P: Changes in the proportions of histone $H1^0$ subtypes in brain cortical neurons. FEBS Lett 1987;210:161–164.

17 Zlatanova JS: Synthesis of histone $H1^0$ is not inhibited in hydroxyurea treated friend cells. FEBS Lett 1980;112:199–202.

18 Chabanas A, Khoury E, Goeltz P, Froussard P, Gjerset R, Dod B, Eisen H, Lawrence JJ: Effects of butyric acid on cell cycle regulation and induction of histone $H1^0$ in mouse cells and tissue culture. J Mol Biol 1985;183:141–151.

19 Pehrson J, Cole RD: Histone $H1^0$ accumulates in growth inhibited cultured cells. Nature 1980;285:43–45.

20 Hall JM, Davis C, Cole RD: Induction of $H1^0$ differs with different treatments among different cell lines. FEBS Lett 1985;189:92–96.

21 Pieler C, Adolf GR, Swetly P: Accumulation of histone $H1^0$ during chemically induced differentiation of murine neuroblastoma cells. Eur J Biochem 1981;115:329–333.

22 Kress H, Tönjes R, Doenecke D: Butyrate induced accumulation of a 2.3 kb polyadenylated $H1^0$ histone mRNA in HeLa cells. Nucleic Acids Res 1986;14:7189–7197.

23 Cayre Y, Raynal MC, Darzynkiewicz Z, Dorner MH: Model for intermediate steps in monocytic differentiation: c-myc, c-fms, and ferritin as markers. Proc Natl Acad Sci USA 1987;84:7619–7623.

24 Barker KA, Newburger PE: Relationships between the cell cycle and the expression of c-myc and transferrin receptor genes during induced myeloid differentiation. Exp Cell Res 1990;186:1–5.

25 Borun TW, Scharff MD, Robbins E: Rapidly labeled, poly-ribosome-associated RNA having the properties of histone messenger. Proc Natl Acad Sci USA 1967;58:1977–1983.

26 Marashi F, Baumbach L, Rickles R, Sierra F, Stein JL, Stein GA: Histone proteins in HeLa S3 cells are synthesized in a cell-cycle stage specific manner. Science 1982; 215:683–685.

27 Unger C, Eibl H: Hexadecylphosphocholine: preclinical and first clinical results of a new antitumor drug. Lipids 1991;26(12):1412–1417.

28 Chirgwin J, Przybyla A, MacDonald R, Rutter W: Isolation of biologically active ribonucleic acid from sources enriched in ribonuclease. Biochem 1979;18:5294–5299.

29 McMaster GH, Carmichael CG: Analysis of single and double-stranded nucleic acids on polyacrylamide and agarose gels by using glyoxal and acridine orange. Proc Natl Acad Sci USA 1977;74:4835–4838.

30 Doenecke D, Tönjes R: Differential distribution of lysine and arginine residues in the closely related histone $H1^0$ and H5. J Mol Biol 1986;187:461–464.

31 Eick S, Nicolai M, Mumberg D, Doenecke D: Human H1 histones: conserved and varied sequence elements in two H1 subtype genes. Eur J Cell Biol 1989;49:110–115.

32 Collins SJ, Gallo RC, Gallagher RE: Continuous growth and differentiation of human myeloid leukemia cells in suspension. Nature 1977;270:347–349.

33 Sandström C, Nilsson K: Establishment and characterization of a human histiocyte lymphoma cell line (U937). Int J Cancer 1976;17:565–577.
34 McKnight S, Tjian R: Transcriptional selectivity of viral genes in mammalian. Cell 1986;46:795–805.
35 Hatzopoulos AK, Schlokat U, Gruss P: Enhancers and other cis-acting regulatory sequences; in Hames BD, and Glover DM (eds): Eukaryotic RNA Synthesis and Processing, Horizons in Molecular Biology Series. Oxford; IRL Press, in press.
36 Lord KA, Hoffmann-Liebermann B, Liebermann DA: Complexity of the immediate early response of myeloid cells to terminal differentiation and growth arrest includes ICAM-1, JunB and histone variants. Oncogene 1990;5:387–396.
37 Sen R, Baltimore D: Inducibility of kappa immunoglobulin enhancer-binding protein NF-kappaB by a posttranslational mechanism. Cell 1986;47:921–928.
38 Leonardo MJ, Baltimore D: NF-kappaB: A pleiotropic mediator of inducible and tissue-specific gene control. Cell 1989;58:227–229.
39 Baeuerle PA, Baltimore D: Activation of DNA-binding activity in an apparently cytoplasmic precursor of the NF-kappaB transcription factor. Cell 1988;53:211–217.
40 Chung J, Sinn E, Reed RR, Leder P: Trans-acting elements modulate expression of the human c-myc gene in Burkitt lymphoma cells. Proc Natl Acad Sci USA 1986;83:7918–7922.
41 Hay N, Bishop JM, Levens D: Regulatory elements that modulate expression of human c-myc. Genes Dev 1987;1:659–671.
42 Remmers EF, Yang JQ, Marcu KB: A negative transcriptional control element located upstream of the murine c-myc gene. EMBO J 1986;5:899–904.
43 Steinbeisser H, Alonso A, Epperlein HH, Trendelenburg MF: Expression of mouse histone $H1^0$ promoter sequence following microinjection into xenopus oocytes and developing embryos. Int J Dev Biol 1989;33:361–368.

Christa Hochhuth, Medizinische Universitätsklinik, Abteilung Hämatologie/Onkologie, Robert-Koch-Strasse 40, D-W–3400 Göttingen (FRG)

Eibl H, Hilgard P, Unger C (eds): Alkylphosphocholines: New Drugs in Cancer Therapy. Prog Exp Tumor Res. Basel, Karger, 1992, vol 34, pp 90–97

Induction of Tumor Cell Differentiation by Alkylphosphocholines: A New Approach for in vitro Screening

H.R. Maurer[a], *P. Hilgard*[b]

[a]Institut für Pharmazie der Freien Universität Berlin, and [b]ASTA Medica AG, Abteilung für Tumorforschung, Frankfurt/Main, FRG

In vitro Screening Assays for Anticancer Drugs Based on Antiproliferation

Most in vitro screening assays for potential anticancer drugs are based on the antiproliferative efficacy of these substances, since uncontrolled cell multiplication is considered an essential property of tumors. Toxic chemicals, however, show similar effects. Therefore, refined methods must be developed and validated in order to distinguish between cytotoxic and cytostatic effects, a difficult task practically. Various *antiproliferative assays* have been developed and used so far, among others: ^{3}H-thymidine uptake assay [Maurer, 1981]; neutral red assay [Borenfreund and Puerner, 1983]; MTT assay [Scudiero et al., 1988]; ATP assay [Garewal et al., 1986]; clonogenic assays [von Hoff, 1988]; dye exclusion assays [Weisenthal et al., 1983]; esterase and phosphatase assays [Martin and Clynes, 1991].

However, some assays may create problems (artefacts), thus restricting the accuracy and specificity of the assays, e.g. up to 14 potential pitfalls of ^{3}H-thymidine techniques to measure cell proliferation have been reported [Maurer, 1981]. In fact, the predictability of such antiproliferative assays is still not satisfactory. Yet they are widely used, not at least due to the introduction of the microtiter plate techniques, facilitating large-scale routine testing.

Induction of Tumor Cell Differentiation

Not all tumors are characterized by increased cell proliferation. Moreover, the effects of many cytostatic drugs cannot be explained by mere cell killing in all cases. Yet many tumor cells show a defective differentiation, i.e. terminal differentiation, as seen within normal cells, is blocked at defined stages of the development. Interestingly, various cytostatic drugs and biological response modifiers, vitamins, cytokines, etc. are capable of inducing tumor cells to differentiate, eventually reversing the malignant phenotype [Bloch, 1984; Freshney, 1985; Cheson et al., 1986; Sachs, 1987]: *Induction of tumor cell differentiation* has since been proposed and studied as a new approach for cancer treatment [Reiss et al., 1986; Marks et al., 1987], also in combination with various drugs [Waxman et al., 1986]. In consequence, prescreening for differentiation-inducing compounds has been applied to select the appropriate compounds to be used for therapy [Lotem et al., 1985]. Moreover, we have found that screening for potential cytostatic cisplatinum derivatives can be improved by combining differentiation assays with clonogenic assays in different leukemic cells [Maurer et al., 1990].

Properties of Cytostatic Alkylphospholipids

Alkyllysophospholipids (ALP) and alkylphosphocholines (APC) are new classes of antineoplastic agents. They were originally derived from the naturally occurring 2-lysophosphatidylcholine [Munder et al., 1976]. This compound and structural analogues are known to accumulate in cell membranes where they exert their cytostatic action [Noseda et al., 1989]. ALP and APC have been shown to inhibit growth of several tumors both in vitro and in vivo [Munder et al., 1976; Muschiol et al., 1987]. In addition, ALP are anti-invasive [Engebraaten et al., 1991]. Interestingly, in animal models and clinical trials their pharmacological side effect profile is quite different from that of most cytostatic drugs [Muschiol et al., 1987; Hilgard et al., 1988]: Virtually no harmful effects on the bone marrow (no myelo- and immunosuppression) and on other actively regenerating tissues; major side effects: hemolysis and gastrointestinal disorders.

One of the most promising APC is hexadecylphosphocholine (HePC = D 18506) [Eibl and Unger, 1990; Hilgard, 1990]. HePC showed in vivo a highly selective antineoplastic activity against autochthonous tumors (DMBA- and MNU- induced rat mammary tumors), while P 388 and L 1210 leukemias, widely used tumor models for cytostatic drug screening, proved to be resistant [Hilgard et al., 1988].

It is not surprising that ALP and APC were found to be differentiation-inducing agents, too. Ten years ago, Honma et al. [1981] were the first to show such effects on human and murine myeloid leukemias in vitro. Later Hochhuth et al. [1990] found that ET-18-OCH_3 and HePC are capable to induce histone synthesis within a few hours as an early

Table 1. Effects of hexadecylphosphocholine (D 18506) on various cells in clonogenic microassays using glass capillaries according to Maurer and Echarti [1990]: IC_{50} = inhibitory concentration at 50% of the control

Cell line	IC_{50}, μM	Range of IC_{50}, μM
P 388	>100.0	>100, >100. >100
L 1210	47.0 ± 8	39, 55, 97
B 16	29.0 ± 1	28, 30
KB	2.7 ± 0.5	1.6, 2.7, 3.0
HL-60	14.5 ± 2	16, 13, 25
Mouse GM-CFC	>100.0	>100, >100, >100

marker of monocytic differentiation of U 937 cells, a human promonocytic leukemia cell line.

When the DMBA tumor, regressing in rats on HePC treatment, was studied histologically over the period of therapy, it was found that already 3 days after HePC administration the mitotic rate decreased and signs of epithelial differentiation appeared [Hilgard et al., 1989]. These changes as well as the reduction of tumor mass by apoptosis became gradually more prominent. Moreover, the absolute amount of tumor stroma increased and original signs of invasiveness at the tumor periphery gradually disappeared.

The KB Tumor for Alkylphospholipid Screening

Moreover, in vitro HePC selectively inhibited KB- and HL-60 colony formation, while other tumor cells (P 388, L 1210 and B 16) were only affected at higher concentrations, normal murine GM-CFC not at all (table 1) [Voegeli et al., 1991]. Our original finding [Echarti and Maurer, 1989] that the human KB squamous cell carcinoma is highly sensitive in vitro to HePC and other phospholipid analogues was also confirmed in vivo on nude mice. The KB tumor is since a specific and preferred tumor model in vitro and in vivo for alkylphospholipid screening.

The conspicuous tumor selectivity in vivo is thus reflected by the in vitro data, excluding general cytotoxicity as a major factor for tumor regression. Therefore, processes other than cytotoxicity must be operating to explain the in vivo efficacy of HePC. Indeed, the histological demonstration of the in vivo induction of differentiation, as mentioned above, was parallelled by our in vitro findings with leukemic cells: HePC induced the human promyelocytic HL-60 cells and the murine myeloid M1-T22 cells to differentiate along the monocytic pathway, as demonstrated by morphology, nonspecific esterase induction, nitroblue tetrazolium (NBT) reduction

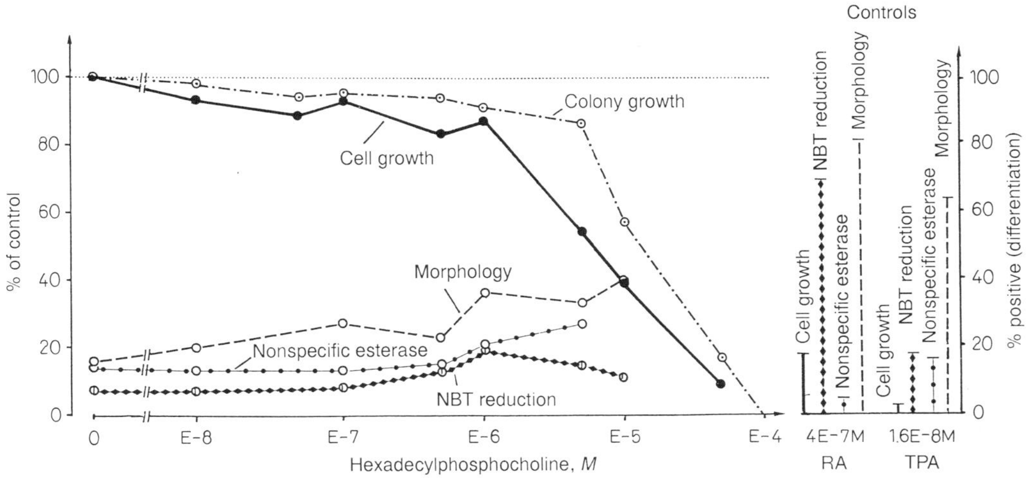

Fig. 1. Effects of hexadecylphosphocholine (D18506) on HL-60 cell proliferation and differentiation.

and phagocytosis, respectively (fig. 1). The human K 562 chronic myelogenous leukemia cell line, now erythroid differentiating, was not induced, however. Remarkably, HePC showed its differentiation-inducing effects at low concentrations that do not reduce cell and colony growth (10^{-6} *M*; fig. 1).

Properties of the Potent Alkylphosphocholine M 71

Recently, M 71 was selected from a large panel of synthetic APC due to its favorable pharmacological and toxicological properties relative to HePC in experimental animal models. Again, this APC showed a strong selective inhibitory activity on KB tumor cells, yet about 10-fold higher than that of HePC. This is reflected by the IC_{50}-ratios of murine GM-CFC/KB: HePC > 30, M 71 > 300. The following IC_{50} values (μM) were determined by clonogenic microassays for M 71: KB 0.35, HL-60 2.3, K 562 ≫ 5, murine GM-CFC ≫ 5.3. In parallel experiments cisplatinum, a cytotoxic anticancer agent, revealed the following IC_{50} values (μM): KB 0.17, HL-60 0.25, K 562 0.18, murine GM-CFC 0.3. It is quite evident from these data that (i) the mode of action of both cytostatics must be different; (ii) the cytotoxicity of cisplatinum does not apply for the APC M 71; (iii) the inhibitory activity of M 71 approaches almost that of cisplatinum for KB cells.

Table 2. Effects of the cytostatic alkylphosphocholine M 71 on differentiation markers of HL-60 cells (human promyelocytic leukemia)

Concentration μM	Differentiated cells (morphology), (+)%	NBT reduced, (+)%	Nonspecific esterase, (+)%
0	4 ± 1.0	9 ± 4.2	4 ± 1.4
1.0	19 ± 9.5	29 ± 14.8	35 ± 0.7
2.5	49 ± 2.1	–	–
5.0	75 ± 1.0	76 ± 19.1	59 ± 2.1

When we tested M 71 in our differentiation assays on leukemic cells in vitro, we found that HL-60 cells reacted up to 75% by both morphology and NBT reduction and 60% by nonspecific esterase staining at 5 μM, values never reached by HePC at that concentration (table 2). This observation was corroborated by dose-response studies showing that as little as 1 μM of M 71 was capable of inducing monocytic cell differentiation in HL-60 cells. In contrast to HePC, M 71 can also induce K 562 cells to produce benzidine-positive cells (20% by 5 μM), supporting the notion that M 71 is a more potent inducer than HePC. Interestingly, K 562 cell colony formation is not inhibited by 5 μM of M 71.

Mode of Action of Alkylphospholipids

The precise mechanisms of action of ALP and APC on the molecular level are still unknown. Increasing evidence suggests that the tumor cell membrane is a significant target for many cytostatic drugs, interfering with signal transduction [Tritton and Hickmann, 1990; Grunicke, 1991]. ALP and APC cause membrane damage in leukemic cells [Noseda et al., 1989] and inhibit protein kinase C [Helfman et al., 1983; Grunicke et al., 1990]. Moreover, Kosano and Takatani (1988) have shown that ALP can reduce EGF binding in human breast cancer cell lines. This suggests the idea that the tumor selectivity of such compounds may be related to the growth factor receptor level of the tumor cell membranes: current experiments will contribute to answer relevant questions.

Differentiation Assays for Alkylphospholipid Screening

It appears from the results reported above that the 'success' of cytostatic drug screening is greatly dependent on the sensitivity of the tumor cell selected for testing. The common L 1210 and P 388 leukemic cells, widely used for cytostatic drug screening, are not suitable for these compounds. Their sole use may well produce false-negative results. Hence our finding of the sensitive KB tumor is substantial for ALP and APC screening. While a highly sensitive tumor model may not be available under certain circumstances, it is advisable to include, into the screening strategy, assay systems other than those based on the endpoint 'antiproliferation'. Indeed, induction of differentiation appears to be an important characteristic phenomen by which distinct cytostatic drugs exert their effects. Interestingly, these effects are seen at concentrations that do not cause immediate inhibition of cell proliferation. Yet induction of differentiation may well lead to inhibition of cell growth and, consequently, to programmed cell death (apoptosis). It follows that the kinetics of the effects must be studied and analyzed in detail at various concentration levels.

Another matter of concern is the choice of the differentiation assays. The endpoints of such assays may be related to the phenotype (morphology), gene expression (biochemical markers) or physiology (functional markers of the cells to be tested). Generally, only several assays of different endpoints will guarantee convincing results. Moreover, in addition to leukemic test cells, carcinoma test cells should be adopted to demonstrate the inducing capacity of drugs. Finally, differentiation assays should be developed and validated such that they can be quickly performed, on a routine basis, by applying microtiter plate techniques.

In *conclusion*, in vitro screening for potential cytostatic ALP and APC should not solely rely upon clonogenic assays with 'antiproliferation' as endpoint, but should include assays with other endpoints such as 'differentiation induction' in order to minimize false-negative results. However, more data are required to demonstrate the general validity of this strategy for broad screening prior to animal and clinical studies.

References

Bloch, A.: Induced cell differentiation in cancer therapy. Cancer Treat. Rep. *68:* 199–205 (1984).

Borenfreund, E.; Puerner, J. A.: A simple quantitative procedure using monolayer cultures for cytotoxicity assays. J. Tissue Culture Meth. *9:* 7–9 (1983).

Cheson, B. D.; Jasperse, D. M.; Chun, H. G.: Friedman, M. A.: Differentiating agents in the treatment of human malignancies. Cancer Treat. Rep. *13:* 129–145 (1986).

Echarti, C.; Maurer, H. R.: The KB tumor for screening of alkylphospholipids (unpublished results, 1989).

Eibl H, Unger C: Hexadecylphosphocholine, a new and selective antitumor drug. Cancer Treat Rev *17:* 233–242 (1990).

Engebraaten, O.; Bjerkvig, R.; Berens, M. E.: Effect of alkyl-lysopholipid on glioblastoma cell invasioninto fetal rat brain tissue in vitro. Cancer Res. *51:* 1713–1719 (1991).

Freshney, R. I.: Induction of differentiation in neoplastic cells. Anticancer Res. *5:* 111–130 (1985).

Garewals, H. S.; Ahmann, F. R.; Schifman, R. B.; Celniker, A.: ATP Assay: Ability to distinguish cytostatic from cytocidal anticancer drug effects. J. natn. Cancer Inst. USA *77:* 1039–1045 (1986).

Grunicke, H.: The cell membrane as a target for cancer chemotherapy. Eur. J. Cancer *27:* 281–284 (1991).

Grunicke, H.; Hofmann, J.; Oberhuber, H.; Überall, F.; Zaknun, J.; Voegeli, R.; Hilgard, P.: Hexadecylphosphocholine inhibits proteinkinase C and depresses the inositol phosphate response in NIH 3T3 fibroblasts. J. Cancer Res. Clin. Oncol. *116:* suppl., part II, p. 889 (1990).

Helfman, D. M.; Barnes, K. C.; Kinkade, J. M.; Vogler, W. R.; Shoji, M.; Kuo, J. F.: Phospholipid-sensitive Ca^{++}-dependent protein phosphorylation system in various types of leukemic cell lines HL-60 and K 562, and its inhibition by alkyl-lysophospholipid. Cancer Res. *43:* 2955–2961 (1983).

Hilgard, P.: New-pharmaceuticals: Miltefosine. Anti-Cancer Drugs *1:* 185 (1990).

Hilgard, P.; Harleman, J. H.; Voegeli, R.; Maurer, H. R.; Echarti, C.; Unger, C.: The antineoplastic activity of hexadecylphosphocholine is associated with tumor cell differentiation. Proc. Am. Ass. Cancer. Res. *30*: 2310 (1989).

Hilgard, P.; Stekar, J.; Voegeli, R.; Harleman, J. H.: Experimental therapeutic studies with miltefosine in rats and mice. Prog exp. Tumor Res., vol. 34, pp. 116–130 (Karger, Basel 1992).

Hilgard, P.; Stekar, J.; Voegeli, R.; Engel, J.; Schumacher, W.; Eibi, H. J.; Unger, C.; Berger, M.: Characterization of the antitumor activity of hexadecylphosphocholine (D 18506). Eur. J. Cancer clin. Oncol. *24:* 1457–1461 (1988).

Hochhut, C.; Berkovic, D.; Eibl, H. J.; Unger, C., Doenecke, D.: Effects of antineoplastic phospholipids on parameters of cell proliferation in U937 cells. J. Cancer Res. clin. Oncol. *116:* 459–466 (1990).

Honma, Y.; Kasukabe, T.; Okabe-Kado, J.; Hozumi, M.; Tushushima, S.; Nomura, H.: Induction of differentiation of cultured human and mouse myeloid cells by aklyllysophospholipids. Cancer Res. *41:* 3211–3216 (1981).

Kosano, H.; Takatani, O.: Reduction of epidermal growth factor binding in human breast cancer cell lines by an aklyllysophospholipid. Cancer Res. *48:* 6033–6036 (1988).

Lotem, J.; Berrebi, A.; Sachs, L.: Screening for induction of differentiation and toxicity to blast cells by chemotherapeutic compounds in human myeloid leukemia. Leukemia Res. *9:* 249–258 (1985).

Marks, P. A.; Sheffery, M.; Rifkind, R.A.: Induction of transformed cells to terminal differentiation and the modulation of gene expression. Cancer. Res. *47:* 659–666 (1987).

Martin, A.; Clynes, M.: Acid phosphatase: Endpoint for in vitro toxicity tests. In vitro *27A:* 183–184 (1991).

Maurer, H. R.: Potential pitfalls of H-thymidine techniques to measure cell proliferation. Cell Tissue Kinet. *14:* 111–120 (1981).

Maurer, H. R.; Echarti, C.: Clonogenic assays for hematopoietic and tumor cells using agar-containing capillaries; in Pollard, Walker, Methods in molecular biology, vol. 5. Animal cell culture, pp. 379–394 (Humana Press, Clifton 1990).

Maurer, H. R.; Echarti, C.; Harleman, J. H.; Hilgard, P.; Voegeli, R.: Induction of tumor cell differentiation by hexadecylphosphocholine in vitro and in vivo. (in preparation 1991).

Maurer, H. R.; Echarti, C.; Voegeli, R.; Pohl, J.; Hilgard, P.: The antitumor activity of D 17872 is associated with tumor cell differentiation. Cancer Chemother. Pharmacol. (submitted, 1991).

Munder, P. G.; Weltzien, H. U.; Modolell, M.: Lysolecithin analogues: A new class of immunopotentiators; in Miescher PA (ed): VIIth Int Symp Immunopathology. Basel, Schwabe, 1976, pp 411–424.

Muschiol, C.; Berger, M. R.; Schuler, B.; Schlerf, H. E.; Garzon, F. T.; Jeller, W. J.; Unger, C.; Eibl, H.; Schmähl, D.: Alkylphosphocholine: Toxicity and anticancer properties. Lipids *22:* 930–934 (1987).

Noseda, A.; White, J.; Godwin, P. L.; Jerome, W. G.; Modest, E. M.: Membrane damage in leukemic cells induced by ether and ester lipids: An electron microscopy study. Exp. mol. Pathol. *50:* 69–83 (1989).

Reiss, M.; Gamba-Vitalo, C.; Sartorelli, A. C.: Induction of tumor cell differentiation as a therapeutic approach: Preclinical models for hematopoietic and solid neoplasms. Cancer Treat. Rep. *70:* 201–218 (1986).

Sachs, L.: Cell differentiation and bypassing of genetic defects in the suppression of malignancy. Cancer Res. *47:* 1981–1986 (1987).

Scudiero, D. A.; Shoemaker, R. H.; Paull, K. D.; Monks, A.; Thierney, S.; Nofziger, Th. H.; Currens, M. J.; Seniff. D.; Boyd, M. R.: Evaluation of a soluble tetrazolium/formazan assay for cell growth and drug sensitivity in culture using human and other tumor cell lines. Cancer Res. *48:* 4827–4833 (1988).

Tritton, T. R.; Hickman, J. A.: How to kill cancer cells: Membranes and cell signaling as targets in cancer chemotherapy. Cancer Cells *2*: 95–105 (1990).

Voegeli, R.; Echarti, C.; Junge, K.; Maurer, H. R.; Hilgard, P.: Correlation between the action of antitumor drugs in colony-forming assays in vitro and the acute toxicity and antitumor activity in vivo (in press, 1990).

Voegeli, R.; Echarti, C.; Maurer, H. R.; Eibl, H. J.; Unger, C.; Stekar, J.; Hilgard, P.: Selective activity of hexadecylphosphocholine against tumor cells in vitro is paralleled by its anti-tumor activity in vivo. Eur. J. Cancer clin. Oncol. (in press, 1991).

Von Hoff, D. D.: Human tumor cloning assays: Applications in clinical oncology and new antineoplastic development. Cancer Metastasis Rev. *7:* 357–371 (1988).

Waxman, S.: The importance of the induction of gene expression and differentiation by cytotoxic chemotherapy. Cancer Invest. *6:* 747–753 (1988).

Waxman, S.; Scher, W.; Scher, B. M.: Basic principles for utilizing combination differentiation agents. Cancer Detection Prevention *9:* 395–407 (1986).

Weisenthal, L. M.: Marsden, J. A.; Dill, P. L.; Macaluse, C. K.: A novel dye exclusion method for testing in vitro chemo-sensitivity of human tumors. Cancer Res. *43:* 749 (1983).

H.R. Maurer, Institut für Pharmazie der Freien Universität Berlin,
Kelchstrasse 31, D-W–1000 Berlin 41 (FRG)

Animal Experiments

Eibl H, Hilgard P, Unger C (eds): Alkylphosphocholines: New Drugs in Cancer Therapy. Prog Exp Tumor Res. Basel, Karger, 1992, vol 34, pp 98–115

Antitumor Activity of Alkylphosphocholines and Analogues in Methylnitrosourea-Induced Rat Mammary Carcinomas

M.R. Berger[a], *P. Yanapirut*[a], *M. Reinhardt*[a], *T. Klenner*[a], *H.R. Scherf*[a], *H.H. Schmeiser*[a], *H. Eibl*[b]

[a]Institute of Toxicology and Chemotherapy, German Cancer Research Center, Heidelberg, and [b]Max-Planck-Institut für biophysikalische Chemie, Göttingen, FRG

Introduction

Autochthonous, chemically induced rat mammary carcinomas are widely used models for the evaluation of compounds interfering with hormonal cell signals derived from the hypothalamic/pituitary/gonadal axis. In comparison with 7,12-dimethylbenz[a]anthracene-induced mammary carcinomas, which are mainly dependent on estrogens, methylnitrosourea (MNU)-induced mammary carcinomas are dependent on estrogens and progestins. In general, the latter tumor model can hardly be influenced by conventional cytotoxic agents (1). Interestingly, alkyllysophospholipids such as 1-O-octadecyl-2-O-methyl-rac-glycero-3-phosphocholine and 1-hexadecylmercapto-2-methoxymethyl-rac-propyl-3-phosphocholine have been shown to exert a consistent anticancer effect in MNU-induced mammary carcinoma [2, 3]. This observation was the basis for using the model to assess the activity of alkylphosphocholines, a new group of anticancer agents which are formally derived from alkyllysophospholipids by relinquishing the glycerol backbone and attaching an alkyl chain directly to the phosphocholine moiety [4]. It could be shown that these agents follow strict dose-response and structure-activity relationships in the model used, with hexadecylphosphocholine (HePC) being the most interesting compound of the first-generation congeners [5]. Subsequently, it was found that topical therapy of patients having locally recurrent breast cancer with HePC is a successful mode of treatment against this disease [6]. Thus, the preclinical results with the MNU model were predictive for the human situation. In view of the fact that a series of often used quickly growing transplanted tumors failed to indicate antineoplastic activity of the alkylphosphocholines, the selective sensitivity and predictivity of the MNU model were reasons to extend experiments on the yet unknown mechanism of action of alkylphosphocholines. Our recent results are summarized in this article.

Material and Methods

Chemicals

Alkylphosphocholines were obtained from H. Eibl (Max Planck Institute of Biophysical Chemistry, Göttingen) and Asta Pharma (Frankfurt, FRG). All compounds were checked for purity by thin-layer chromatography (TLC) and elemental analysis and were dissolved in water at appropriate concentrations shortly before use. MNU was synthesized by M. Wiessler (Institute of Toxicology and Chemotherapy, German Cancer Research Center, Heidelberg, FRG) and dissolved at 1% shortly before use in Sörensen buffer, pH 6, and distilled water (25/75, v/v).

Animals and Tumor Induction

Female SD and Fisher rats were purchased from the Zentralinstitut für Versuchstierzucht (Hannover, FRG) and maintained under conventional conditions (temperature 22 ± 2 °C, relative humidity $55 \pm 10\%$, dark-light rhythm of 12 h). Altromin pellets and water were given ad libitum.

Autochthonous, hormone-dependent mammary carcinomas were induced by MNU in female SD rats as described previously [1]. Rats with a total tumor volume >0.8 cm^3 were randomly allocated to experimental groups, therapy started immediately thereafter for a period of 5–10 weeks.

Ovariectomy was performed according to standard surgical procedures in rats which fulfilled the criterion of eligibility for therapy (tumor volume); these animals remained either untreated as ovariectomized controls or received additional chemotherapeutic therapy. Moribund animals were killed and dissected; mammary carcinomas were excised, weighed, fixed in a 7% formalin solution and examined histologically. Tumors collected for DNA analysis were shock frozen in liquid nitrogen and processed to histology and DNA analysis. Similarly, tumors, liver and plasma taken from rats for cAMP determination were frozen in liquid nitrogen and kept at −70% °C until analysis.

Evaluation of Antitumor Activity

The tumor volume in treated and untreated rats was recorded during the whole experimental period. The median tumor volumes of groups were expressed in percent of untreated controls (T/C%).

Tumor Transplantation

Following induction of Fisher rats with MNU a mammary carcinoma was excised and parts of this tumor were either processed to histology or used for tumor transplantation. To that purpose the remaining tumor was minced with scissors in RPMI medium enriched with growth factors and the resulting small tumor pieces were mashed through a metal sieve. 100 mg of the resulting tumor pulp were inoculated s.c. to syngeneic female rats. When the tumor had grown to a size of 1–2 cm^3 the procedure was repeated. At passages 5 and 10, forty rats were used, respectively, for determination of sensitivity to HePC.

DNA Analysis

Individual tumors were collected, quickly frozen in liquid nitrogen, and stored at −80 °C until DNA isolation. High-molecular-weight DNA was prepared according to standard methods.

Polymerase Chain Reaction

For *in vitro* amplification of DNA, the method described by Saiki et al. [7] was used. One microgram of DNA was amplified in a reaction mixture with a total volume of 100 μl containing 67 m*M* Tris-HCl (pH 8.8), 16.6 m*M* ammonium sulfate, 6.7 m*M* magnesium chloride, 10 m*M* 2-mercaptoethanol, 6.7 μ*M* EDTA, 1.5 m*M* dATP, dCTP, dGTP, and dTTP, 170 μg/ml bovine serum albumin, and 1 μ*M* concentration of each primer. The incubation mixture was overlaid with a few drops of paraffin oil, and the DNA was denatured at 94 °C for 10 min. Thereafter, the polymerase reaction was started by addition of 2 U of TAQ polymerase (Perkin Elmer-Cetus). Usually, we performed 35 cycles of amplification, each cycle consisting of denaturation (94 °C, 1.5 min), primer annealing (37 °C, 1.5 min), and chain elongation (72 °C, 2 min). At the end of the last cycle, the polymerase reaction was allowed to complete at 72 °C for 5 min. Control incubations were performed without DNA.

The primers used for amplification of the region around codon 12 of the H-*ras* gene were 5′-ACCCCTGTAGAAGCGATGAC-3′ and 5′-CACAAAATGGTTCTGGATCA-3′. Amplifications yielded a 102-bp DNA fragment, which was visualized on 2% agarose gels containing 0.5 μg ethidium bromide/ml as a specific band.

Oligonucleotide Hybridization

Aliquots of the amplification mixture were diluted with 10 m*M* TrisHCl/1 m*M* EDTA in the range between 1:5 and 1:20. Twenty microliters of the solution were spotted onto a nylon membrane (Bio Trace RP; Gelman Sciences, Ann Arbor, Mich., USA), and the DNA was fixed by baking the membrane at 80 °C for 2–3 h. The membranes were prehybridized overnight at 50 °C in 5 × SSPE containing 0.3% SDS and 500 μg/ml of denatured salmon sperm DNA. Hybridization with oligonucleotide probes end labeled (5′) with ^{32}P was performed overnight at 50 °C in the same solution. Thereafter, the membranes were washed in 2 × SSPE-0.1% SDS at room temperature, followed by a stringent wash in 5 × SSPE-0.1% SDS for 10 min. After a final wash in 2 × SSPE-1% SDS at room temperature, the membranes were exposed to Kodak X-ray films overnight at −70 °C using intensifying screens. Oligonucleotide probes used were 19-mers from NEN Division, Dreieich, FRG. A complete list of oligomers used for our analyses of H-*ras* gene mutations is available on request.

Determination of cAMP Levels

All preparations were carried out at a maximum temperature of 0 °C according to the following procedure: tissue fragments were roughly crushed between precooled aluminium blocks (−196 °C) and pulverized twice with a ball mill (type Micro-Dismembrator II, Braun, Melsungen, FRG) for 45 s each with a 5-min cooling break. Approximately 100 mg tissue powder was homogenized in 2 ml ice-cold 10% trichloroacetic acid (TCA), using an all-glass homogenizer (Dounce). After centrifugation of the homogenate at 2,500 *g*, the supernatant was extracted three times with 5 ml of water-saturated ether. The aqueous extracts were lyophylized, redissolved in assay buffer and stored at −70 °C. All samples were diluted one to one with buffer (Tris-HCl buffer pH 7.4 with EDTA) and directly applied in the assay. Cyclic AMP levels were determined with a kit (Immundiagnostik, Darmstadt, FRG) which is based on competitive protein binding analysis as described earlier [8]. Shortly, ^{3}H-labeled cAMP was used as a tracer, cAMP-binding protein was extracted from bovine cardiac muscle, 50-μl samples were incubated at 4 °C with 20 μl tracer (specific activity: 165 KBq/2 ml) and 20 μl binding protein (concentration: 3 mg/l). After 3 h 100 μl charcoal (1%, w/v) was added and again incubated for 5 min. This mixture was centrifuged for 10 min at 1,800 *g* and 100 μl of the supernatant were added to 2 ml scintillation cocktail (Ready Safe, Beckmann, FRG) and counted for 2 min in a liquid scintillation counter (2000 CA Tricarb, Packard). The cAMP levels of the samples were derived from a standard curve which has a measuring range of 3.8–320 nmol/L (all reagents in Tris-HCl buffer pH 7.4 with EDTA and 20 g/l protein supplement) using a software for evaluation of RIAs (Radek Labordiagnostik). Values were adjusted to 100 mg of wet tissue weight.

Results

The anticancer efficacy of regular administration (five times weekly for 5 weeks) of tetradecylphosphocholine (TePC), HePC and octadecylphosphocholine (OcPC) is given in table 1 and figures 1–3. TePC shows a

Table 1. Antineoplastic efficacy of TePC, HePC and OcPC in MNU- induced rat mammary carcinoma

Compound	Dose[1] μmol/kg	Number of animals	T/C%[2]	ILS[3]	% body weight change[4]	% mortality[5]
Control*	–	20	100	–	12	5
Contol**	–	20	100	–	6	25
TePC**	172	15	42	51	0	13
TePC*	115	15	72	5	9	13
TePC*	77	15	74	12	7	13
TePC*	51	15	92	–5	–5	20
HePC*	115	15	toxic	–69	toxic	93
HePC*	77	15	2	26	–4	40
HePC*	51	15	13	23	2	0
HePC**	34	15	39	38	8	7
HePC**	15	10	28	24	8	10
HePC**	7	10	74	16	2	10
OcPC*	115	15	toxic	–76	toxic	100
OcPC*	77	15	0.6	–72	–13	80
OcPC*	51	15	2	26	–7	33
OcPC**	34	15	6	53	0	0
OcPC**	15	15	23	53	4	0
OcPC**	7	10	74	2	2	20

* = Experiment A; ** = experiment B.
[1]Administration p.o. 5 × /week for 5 weeks.
[2]Quotient of median tumor volumes of treated and control groups × 100; evaluation at week 6.
[3]Percent increase in life span over untreated controls.
[4]Weight difference at week 6 in percent of initial body weight.
[5]Evaluation at week 6.

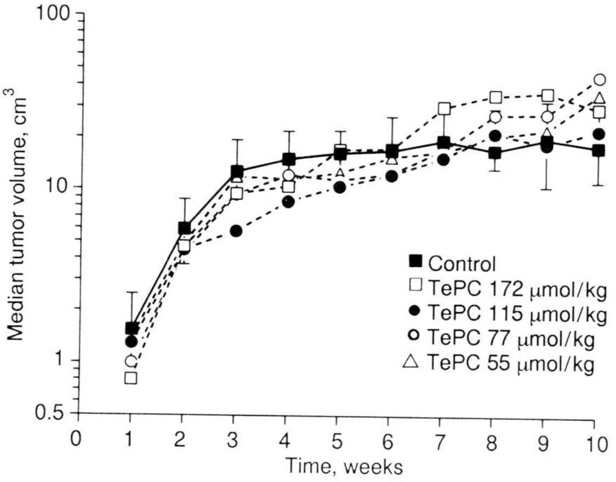

Fig. 1. Antineoplastic efficacy of regular administration of TePC for 5 weeks in MNU-induced rat mammary carcinoma.

moderate activity without signs of toxicity at the highest dosage used (172 μmol/kg; fig. 1) which was also documented by a 51% increased life span. Compared to this high dosage the optimal dose of HePC was about threefold lower (51 μmol/kg) and caused a threefold higher tumor growth inhibition (13 *versus* 42 T/C%) at comparable toxicity. The two dosages of HePC higher than the optimal dose were poorly tolerated and caused high mortality, although the animals surviving a dose of 77 μmol/kg experienced a marked reduction of the median tumor volume (fig. 2, top) and a considerable increase in life span. In the investigated dose range a clear dose-response relationship was obtained with negligible anticancer activity following the lowest dose (7 μmol/kg). Compared to HePC its analog OcPC was somewhat more toxic on an equimolar base, the optimal dose being 34 μmol/kg (fig. 3). Higher dosages of OcPC caused intolerable toxicity but the two lower dosages used (15 and 7 μmol/kg) were not more effective than the respective dose levels of HePC thus indicating a narrower therapeutic index of OcPC compared to HePC. From these results it is concluded that HePC is the most promising agent of this series which, therefore, was to be investigated in more detail. The optimal dose of HePC was reinvestigated and, for two groups of animals, the duration of administration was prolonged to 10 weeks with either a constant dose administered for the whole period (51 μmol/kg, fig. 4, top) or a double dose administered every second week (115 μmol/kg, fig. 4, bottom). As is evident from figures 2 and 4 (top) cessation of therapy after 5 weeks induced regrowth of mammary carcinomas within 1 week, the slope of the

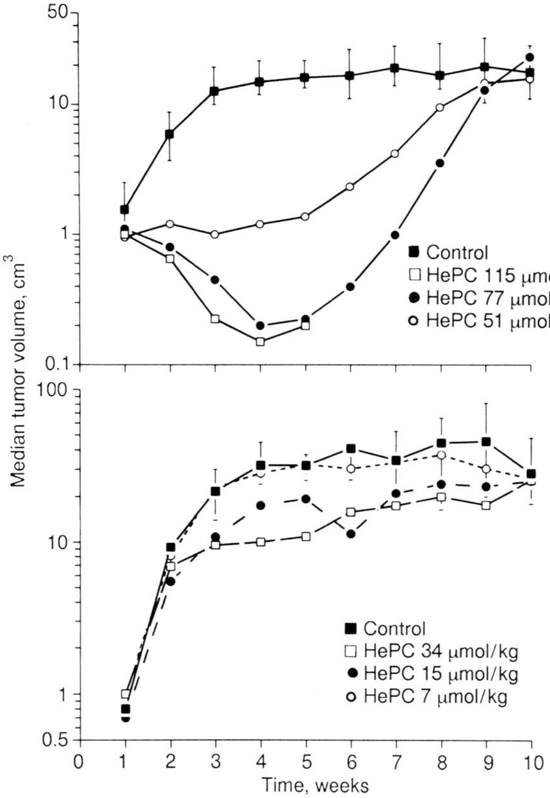

Fig. 2. Antineoplastic efficacy of regular administration of HePC for 5 weeks in MNU-induced rat mammary carcinoma.

increase in tumor volume being similar to that of controls. When the therapy was continued, however, no significant increase in tumor volume was observed during this time (fig. 4, top). Intermittent therapy with high dosages was also effective, with the last treatment still inducing a decrease in tumor volume (fig. 4, bottom) and thus indicating no development of resistance within that time interval.

The observation that regular therapy administered every second week can prevent outgrowth of tumors was basis for assuming a relatively long half-life of HePC and introducing the use of pulse doses, administered once weekly. Additionally, the question was investigated whether HePC acts equally effective in hormone-dependent and hormone-independent mammary carcinomas. The design of this experimental series is given in table 2.

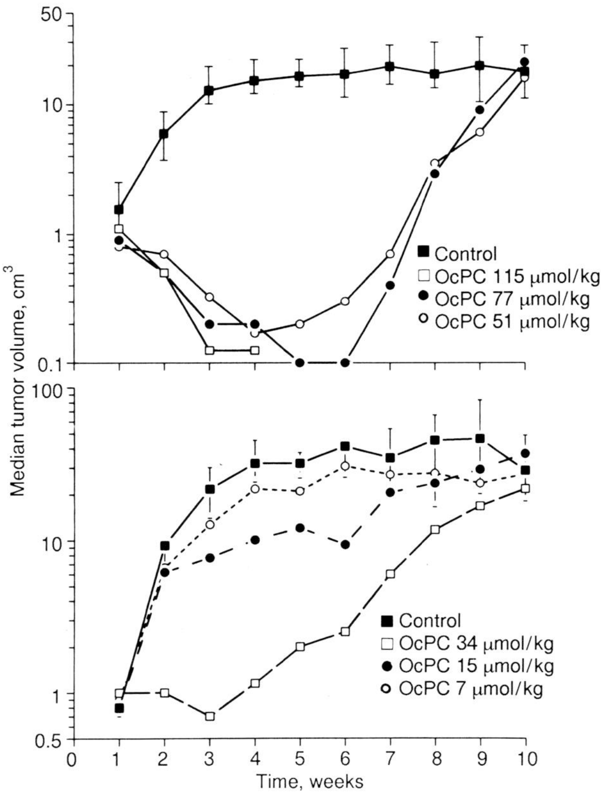

Fig. 3. Antineoplastic efficacy of regular administration of OcPC for 5 weeks in MNU-induced rat mammary carcinoma.

Interestingly, a pulse dose of 150 mg/kg (368 μmol/kg) which is about equivalent to administering five times 77 μmol/kg was better tolerated than regular administration and found to effectively inhibit tumor growth (fig. 5, top). Although no decrease in median tumor volume was evident as seen following administration of five times 77 μmol/kg (fig. 2, top) the T/C% value following 368 μmol/kg once weekly (4; see table 2) was identical to that following 77 μmol/kg five times weekly (2; see table 1) after 5 weeks of treatment. Similarly, pulse doses of 100 mg/kg (245 μmol/kg) and 67 mg/kg (164 μmol/kg) were equally active as dosing five times 51 and 34 μmol/kg, respectively.

Ovariectomy alone caused a mild effect on the growth of MNU-induced mammary carcinomas, as indicated by 49 T/C% at week 6 after

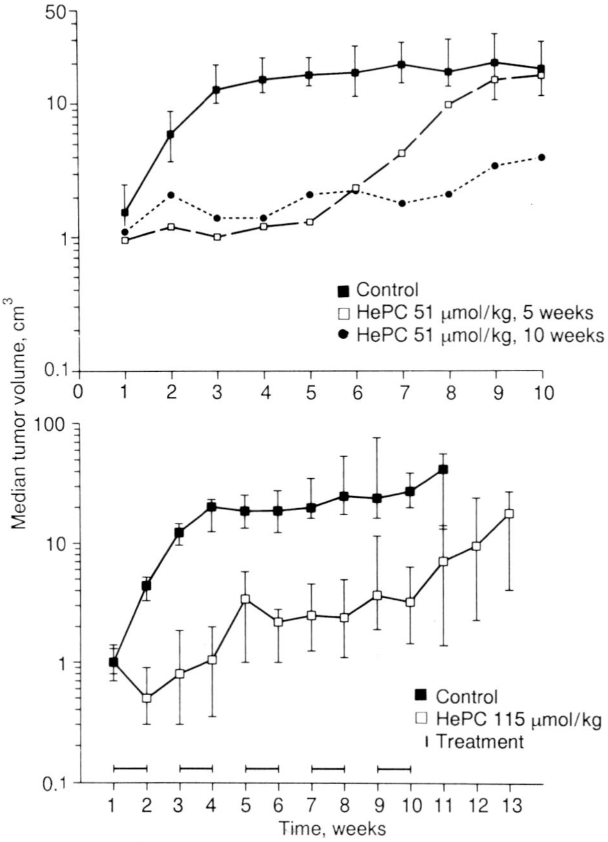

Fig. 4. Administration of HePC to MNU-induced Sprague-Dawley rats bearing mammary carcinoma: Antineoplastic efficacy of regular administration for 5 or 10 weeks (top) and of intermittent administration every second week (bottom).

castration of the animals (table 2; fig. 5, bottom). Additional treatment with HePC with doses of 150 and 100 mg/kg caused additional tumor growth inhibition corresponding to 1.4 and 10 T/C% values. This demonstrates that HePC is as effective in hormone-independent tumors growing in castrated animals as in hormone-dependent neoplasias growing in intact rats. The increased mortality of castrated animals treated with HePC was due to their higher body weight, which increased as a side effect of the ovariectomy, since the absolute dose per rat was raised with the body weight.

Table 2. Influence of pulse doses of HePC on intact and ovariectomized Sprague-Dawley rats bearing MNU-induced mammary carcinoma

Treatment	Dose[1]		Number of animals	T/C%[2,3]	% body weight change[2,4]	% mortality[2]
	mg/kg	μmol/kg				
Control	–	–	39	100	18	29
HePC	150	368	12	4	4	0
HePC	100	245	14	17	6	6
HePC	67	164	15	34	12	15
Ovariectomy	–	–	18	100 (49)[5]	10	6
Ovariectomy + HePC	150	368	14	1.4	– 2	43
Ovariectomy + HePC	100	245	12	10	– 4	17

[1]Peroral administration once weekly for 5 weeks.
[2]Evaluation at week 6 after start of therapy (week 1).
[3]Quotient of median tumor volumes of treated and control groups × 100.
[4]Body weight change in % of initial body weight.
[5]The value in brackets denotes the T/C% value of ovariectomized rats in relation to untreated controls, which was set to be 100%.

Table 3. Overview of tumors with sensitivity to HePC in vivo

Tumor model	Sensitivity	References
Methylnitrosourea-induced rat mammary carcinoma	+	Berger et al., 1987 [5]
3,4-Benzo(a)pyrene-induced rat fibrosarcoma	±	Berger et al., 1987 [5]
Acetoxymethylmethylnitrosamine-induced rat colorectal carcinoma	–	Berger et al., 1987 [5]
Transplanted rat mammary carcinoma TMA1	±	Scherf et al., 1987 [19]
Transplanted rat mammary carcinoma TMA2	±	Scherf et al., 1987 [19]
Transplanted rat mammary carcinosarcoma Walker 256B	–	Muschiol et al., 1987 [20]
Transplanted mouse mammary carcinoma MXT	–	Muschiol et al., 1987 [20]
Dimethylbenzanthracene-induced rat mammary carcinoma	+	Hilgard et al., 1988 [21]
Transplanted mouse leukemia L1210	–	Engel et al., 1988 [22]
Transplanted mouse leukemia P388	–	Engel et al., 1988 [22]
Transplanted KB cell carcinoma	+	Hilgard et al., 1991 [23]
Transplanted rat leukemia L5222	–	Berger et al., 1991 (unpublished results)
Transplanted mammary carcinoma PY(passage 5)	+	Reinhardt et al., 1991 (unpublished results)
Transplanted mammary carcinoma PY(passage 10)	–	Reinhardt et al., 1991 (unpublished results)

The selectivity of the antineoplastic action of HePC is indicated from table 3, which gives an overview on a series of tumors being sensitive or insensitive to HePC. Besides MNU- and DMBA-induced mammary carcinomas only few transplanted tumors have shown sensitivity to treatment

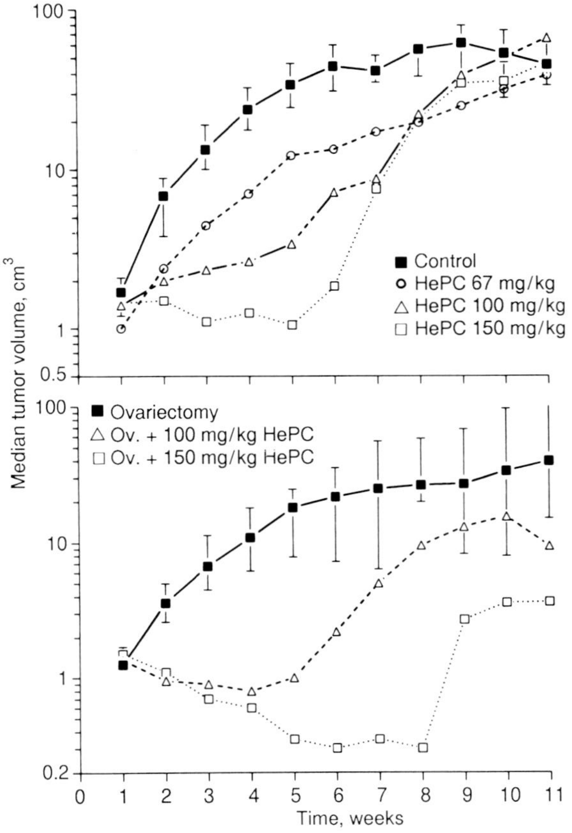

Fig. 5. Antineoplastic efficacy of pulse doses of HePC in intact (top) and ovariectomized (bottom) rats bearing MNU-induced mammary carcinoma.

with HePC such as the KB cell tumor growing in nude mice. In an experiment to study a possible change in sensitivity to HePC with increasing number of passages of a transplanted tumor, a MNU-induced mammary carcinoma was serially transplanted in Fisher rats. When this tumor was treated in the fifth passage (fig. 6, top) a distinct sensitivity to treatment with HePC was evident. During the following passages, however, the mammary carcinoma showed a quicker manifestation time and a more rapid growth following transplantation. Concomitantly with these changes the tumor became resistant to HePC, as shown in figure 6 (bottom).

In an attempt to characterize the action of HePC on MNU-induced mammary carcinoma histological slides of tumors from treated and un-

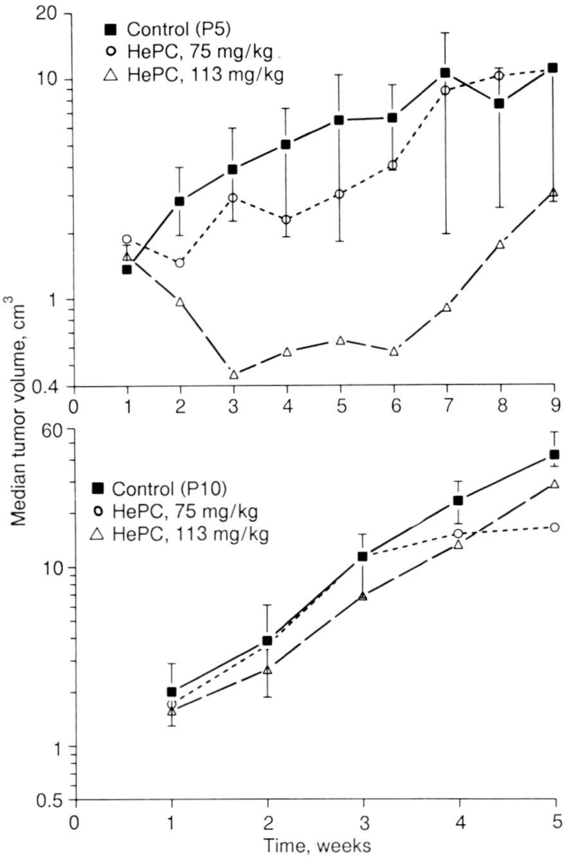

Fig. 6. Influence of pulse doses of HePC on a mammary carcinoma transplanted in Fischer 344 rats in passage 5 (top) and passage 10 (bottom).

treated rats were evaluated. A comparison showed that HePC treatment induced changes in tumor morphology that can be described as follows: reduction of nuclear size and nuclear polychromasia, reduced numbers of epithelium layers, increased stroma reaction, and increased lactation. These observations are consistent with increased differentiation of these tumors induced by HePC (fig. 7).

Sukumar et al. [9] have shown that the induction of mammary carcinomas in rats by MNU involves the malignant activation of the c-Ha-*ras* gene by a point mutation in codon 12. To investigate whether HePC induces changes in the activated codon of treated tumors, codon 12

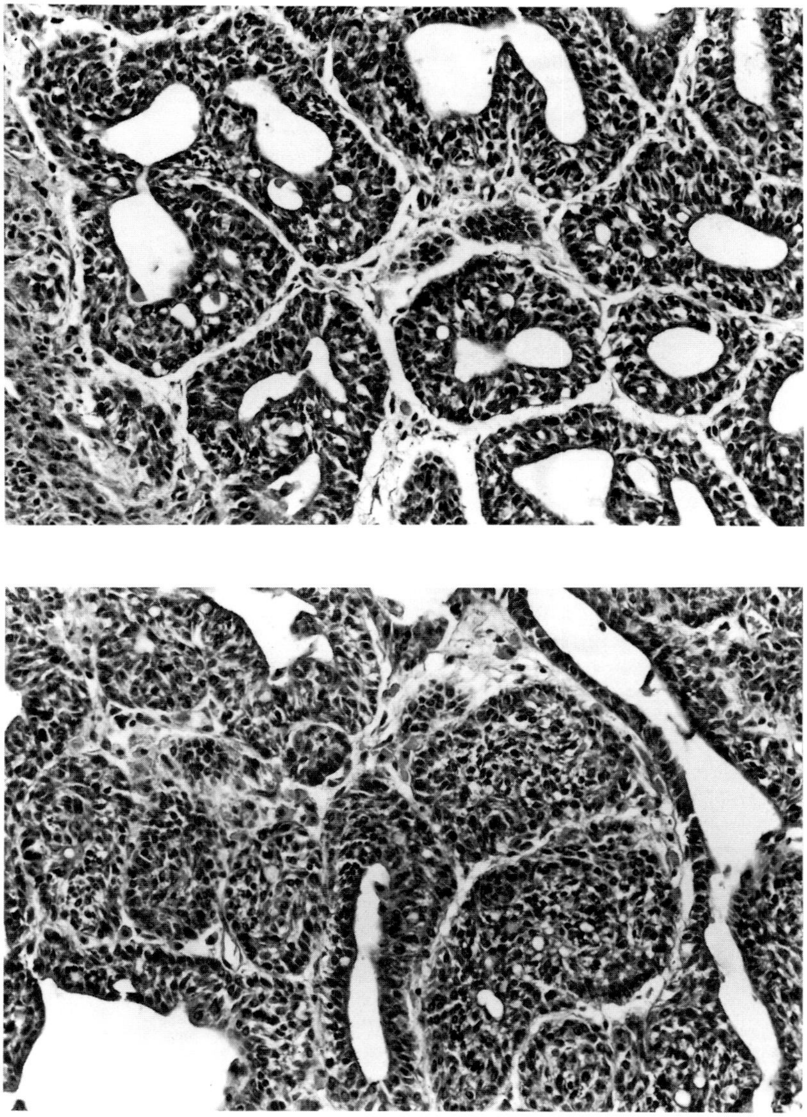

a

b

Fig. 7. Microscopic view of MNU-induced mammary carcinomas which are untreated (*a*, *b*) or treated (*c*, *d*; see p. 110) with HePC.

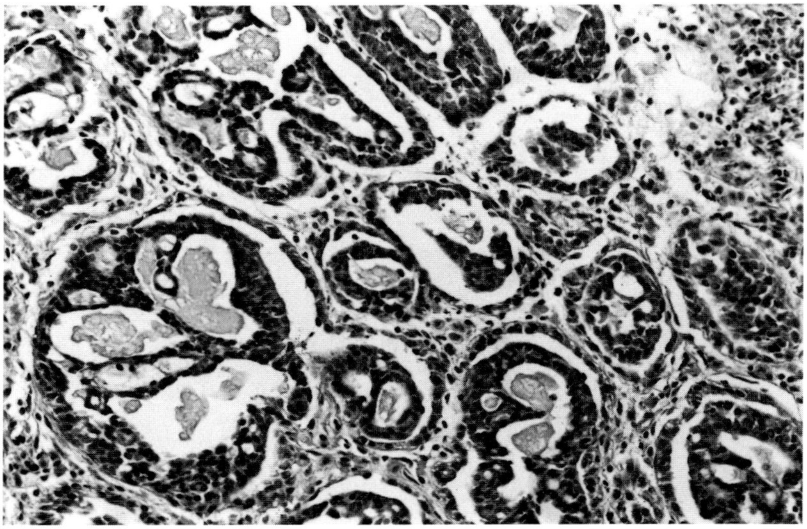

c

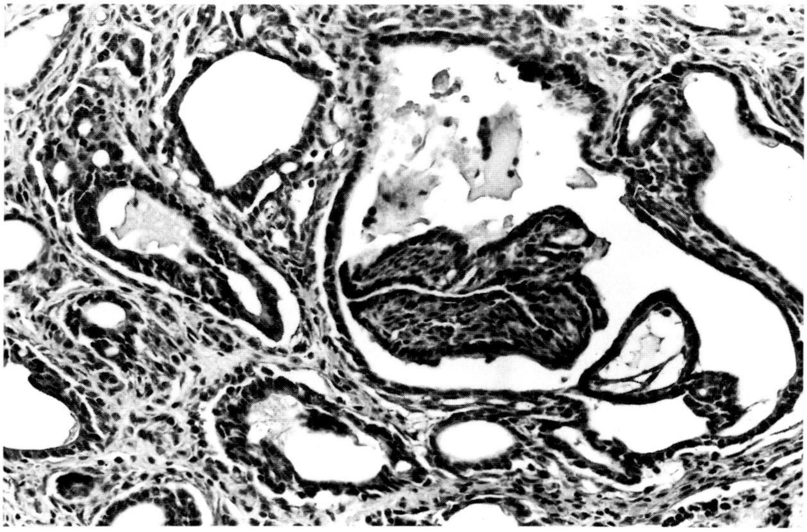

d

Fig. 7. c, d. (For legend see previous page.)

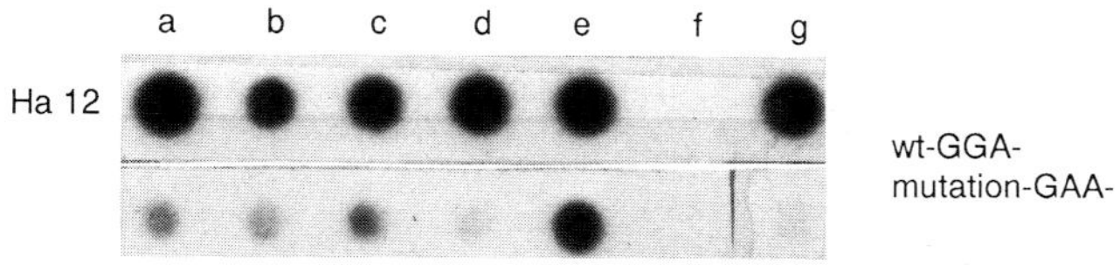

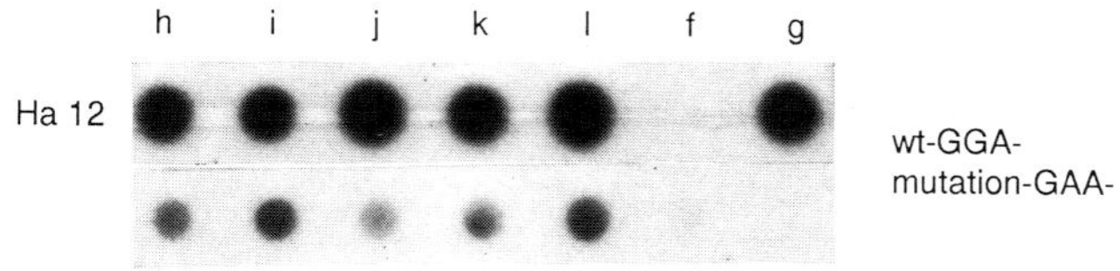

Fig. 8. Hybridization of mutation specific oligomers to *in vitro* amplified DNAs. Approximately 5 ng of DNAs were spotted onto a nylon membrane and hybridized to oligomers representing wild type (*wt*) and mutation specific codon 12 Ha-*ras* sequences. a-e = Tumors untreated; f = control without DNA; g = rat forestomach tumor DNA, with mutation in codon 61 of Ha-*ras* gene; h-l = tumors treated.

of the c-Ha-*ras* oncogene was characterized by the polymerase chain reaction. DNAs from MNU-induced mammary carcinomas of treated and untreated rats were amplified for the region around codon 12 of the c-Ha-*ras* gene, followed by selective oligonucleotide hybridization. The probes used for hybridization included oligomers specific for every possible amino acid substitution at codon 12. A GC → AT transition mutation at the second position of codon 12 of the c-Ha-*ras* gene, resulting in a change of glycine (codon GGA) with glutamic acid (codon GAA) was detected in all untreated and treated MNU-induced mammary carcinomas, as shown in figure 8. The weak signals of the mutated allele (GAA) in figure 8 are possibly due to contaminating normal cells in the analyzed tumor material. The results show that HePC induces no alteration in the activation of the c-Ha-*ras* oncogene.

The influence of 200 mg/kg HePC administered as single dose on the time course of cAMP levels in mammary carcinomas, liver and plasma of MNU-induced tumor-bearing rats is given in table 4. The quickest response of cAMP was found in plasma, showing a twofold rise at 1 h and a subsequent fall to 50% of initial levels 4–16 h following HePC administration. Liver tissue, on the other hand, showed an initial decrease in cAMP

Table 4. Modulation of cAMP levels following p.o. administration of 200 mg/kg HePC to female Sprague-Dawley rats bearing MNU-induced mammary carcinoma.

Time following p.o. treatment	Mean ± SD cAMP level, nmol/100 mg sample		
	mammary carcinoma (n)	liver (n)	plasma (n)
Control	60 ± 6 (6)	73 ± 24 (2)	96 ± 11 (2)
1 h	63 ± 17 (12)	62 ± 6 (3)	193 ± 31 (3)
2 h	77 ± 27 (9)	38 ± 4 (2)	155 ± 68 (3)
4 h	67 ± 34 (14)	66 ± 41 (3)	47 ± 24 (3)
8 h	104 ± 29 (13)	86 ± 8 (3)	48 ± 2 (3)
16 h	120 ± 33 (12)	110 ± 13 (3)	48 ± 24 (3)
24 h	84 ± 24 (9)	66 ± 10 (2)	61 ± 22 (3)
48 h	71 ± 6 (8)	60 ± 8 (3)	38 ± 15 (3)
96 h	114 ± 23 (11)	154 ± 42 (4)	80 ± 34 (4)
192 h	76 ± 27 (6)	110 ± 1 (2)	79 ± 46 (3)

N = Number of probes.

levels at 2 h and subsequently a double peak at 16 and 96 h following treatment. Mammary carcinomas showed no quick response but paralleled the double peak as observed in liver tissue at 16 and 96 h following HePC.

Discussion

The exceptionally high antineoplastic activity of alkylphosphocholines, especially of HePC, against chemically induced mammary carcinomas is striking, since a variety of conventional agents used in the clinic, or interesting new experimental drugs were less effective against this tumor [1, 10, 11]. Alkylphosphocholines even prevail hormonal or hormone-linked anticancer agents which hitherto had shown the highest efficacy in this model [12, 13]. Furthermore, the highly selective action of HePC, which in vivo is active only in few tumor models but not in series of systems, as leukemias P388 and L1210, that have been used for their broad sensitivity to the majority of known anticancer agents, is surprising as well. Both observations, when taken together, suggest a high degree of specificity in the action of HePC against chemically induced mammary carcinomas. If so, an interference with a metabolic pathway vital for the cancerous growth can be assumed. Which pieces of evidence exist to date to support this assumption? By their physical properties alkylphosphocholines target the cell membrane but they interfere with the signal transduction as evidenced by their effect on cAMP rather than lyse cell membranes unspecifically. Similarly, other second messengers could be influenced which then are responsible for inhibition of, e.g. phospholipases C or D and protein kinase C. The latter enzyme was recently found to be inhibited by agents of this group [14, 15]. The physical properties of HePC exclude a direct influence on the DNA of cells and in line with this assumption no alteration in the activating codon of the c-Ha-*ras* oncogene was found in MNU-induced mammary carcinomas (fig. 7). The transcription and translation of DNA is, however, regulated by signal cascades, in which second messengers are involved and consequently the expression of oncogenes under the influence of alkylphosphocholines should be investigated in forthcoming studies. Indeed, first experiments have been reported in which increased synthesis of m-RNA of a histone protein was found in response to exposure to HePC [16]. Indirect influence of HePC on DNA has furthermore been shown in experiments on the cell cycle: Mammary carcinoma cell lines derived from MNU-induced neoplasias were arrested in G1/G0 phase of the cell cycle following exposure to HePC [17]. This stage of the cell cycle is known to be under hormonal control of the body. Both mechanisms, the reduced cell division due to HePC and the arrest in a stage of the cell cycle more prone to intercell regulation could be responsible for the observed increase in differentiation in MNU-induced mammary carcinomas treated with HePC in vivo.

Experiments on structure-activity relationships carried out with a series of congeners of HePC have shown that the anticancer activity of this group is restricted to certain structural properties of the molecule. Parallel with the increasing alkyl chain length toxicity and anticancer efficacy of alkylphosphocholines were stronger. In this series HePC had the highest therapeutic ratio (table 1). Introduction of a double bond into the alkyl chain also has an influence on this parameter. When comparing OcPC (18:O-PC) with octadecenyl-(trans-9,10)-phosphocholine (t-18:1-PC) and octadecenyl-(cis-9,10)-phosphocholine (c-18:1-PC), the therapeutic index of these compounds was found to correspond to a ratio of 18:0-PC/t-18:1-PC/c-18:1-PC = 1/2/4 [3]. Besides the length and configuration of the alkyl chain the phosphocholine moiety was found to be of similar importance. Extension of the distance between the phosphate and the trimethylammonium group to 3, 4 or 6 carbon atoms to generate hexadecylphospho(N,N,N-trimethyl)propanolamine, -butanolamine, and hexanolamine resulted in structures with progressively decreasing anticancer activity [18].

With regard to the model used, it is important to discriminate between sexual hormone-dependent and hormone-independent mechanisms of action. The high antineoplastic activity of HePC in mammary carcinomas growing in ovariectomized rats demonstrates a hormone-independent mechanism of action of this group of compounds. The sensitivity of chemically induced mammary carcinomas could then be based on the fact that they contain a mutated c-Ha-*ras* oncogene which leads to the expression of an altered G-protein located in the membrane and involved in signal transduction. Interference with this pathway in which cellular signals probably are inadequately transduced seems to be a plausible assumption for the mechanism of action of alkylphosphocholines in the model used. The sensitivity of these tumor cells is, however, lost following several passages in inbred animals. This phenomenon could be related to the outgrowth of primarily resistant cells or to the appearance of resistant tumor cells which probably contain more than one lesion leading to cancerous growth.

In summary, alkylphosphocholines are a promising new class of anticancer agents which seem to have a specific mechanism of anticancer action and could be used, therefore, for a more specific therapy than other clinically available agents.

References

1 Berger MR, Habs M, Schmähl D: Non-carcinogenic chemotherapy with a combination of vincristine, methotrexate and 5-fluorouracil (VMF) in rats. Int J Cancer 1983;32:231–236.

2 Berger MR, Munder PG, Schmähl D, Westphal O: Influence of the alkyllysophospholipid ET-18-OCH*3* on methylnitrosourea-induced rat mammary carcinomas. Oncology 1984;41:109–113.

3 Berger MR, Richter H, Seelig MH, Eibl H, Schmähl D: New cytostatics: More activity and less toxicity. Cancer Treat Rev 1990;17:143–154.

4 Eibl H, Unger C: Phospholipide als Antitumormittel: Möglichkeiten einer selektiven Therapie; in Unger C, et al (ed): Die Zellmembran als Angriffspunkt der Tumortherapie. Aktuelle Onkologie. München, Zuckschwerdt, 1987, vol 34, pp 1–18.

5 Berger MR, Muschiol C, Schmähl D, Eibl H: New cytostatics with experimentally different toxic profiles. Cancer Treat Rev 1987;14:307–317.

6 Unger C, Eibl H, Nagel GA, von Heyden HW, Breiser A, Engel J, Stekar J, Peukert M, Hilgard P, Berger MR: Hexadecylphosphocholine in the topical treatment of skin metastases: A phase I trial; in Queisser W, Fiebig HH (eds): New Drugs in Oncology. Contributions to Oncology. Basel, Karger, 1989, vol 37, pp 219–223.

7 Saiki RK, Gelfand DH, Stoffel S, Scharf SJ, Higuchi R, Horn GT, Mullis KB, Ehrlich HA: Primer-directed enzymatic amplification of DNA with a thermostable DNA polymerase. Science 1988;239:487–491.

8 Armbruster FP, Scharfenstein H, Hitzler W, Schmidt-Gayk H: cAMP-Protein-Bindungs-Assay: Proteinabhängigkeit der cAMP-Bindung. Ärztl Lab 1986;32:115–120.

9 Sukumar S, Notario V, Martin-Zanca D, Barbacid M: Induction of mammary carcinomas in rats by nitroso-methylurea involves malignant activation of H-*ras*-1 locus by single point mutations. Nature 1983;306:658–661.

10 Berger MR, Zeller WJ: Interaction of nicotine with anticancer treatment. Klin Wochenschr 1988;66(suppl XI):127–133.

11 Schwendener RA, Fiebig HH, Berger MR, Berger DP: Evaluation of incorporation characteristics of mitoxantrone into unilamellar liposomes and analysis of their pharmacokinetic properties, acute toxcity, and antitumor efficacy. Cancer Chemother Pharmacol 1991;27:429–439.

12 Berger MR, Floride J, Schmähl D, Schreiber J, Eisenbrand G: Estrogen-linked 2-chloroethylnitrosoureas: Anticancer efficacy in MNU-induced rat mammary carcinoma, uterine activity in mice and receptor interactions. Eur J Cancer Clin Oncol 1986;22:1179–1191.

13 Brix HP, Berger MR, Schneider MR, Tang WC, Eisenbrand G: Androgen-linked alkylating agents: Biological activity in methylnitrosourea-induced rat mammary carcinoma. J Cancer Res Clin Oncol 1990;116:538–549.

14 Hofmann J, Ueberall F, Posch L, Maly K, Herrmann DBJ, Grunicke H: Synergistic enhancement of the antiproliferative activity of cis-diamminedichloroplatinum(II) by the ether lipid analogue BM 41440, an inhibitor of protein kinase C. Lipids 1989;24:312–317.

15 Vogler WR, Olson AC, Okamoto S, Shoji M, Kuo JF, Berdel WE: Comparative effects of lipids on leukemic cell lines and protein kinase C (PKC). Proc Am Assoc Cancer Res 1989;30:577.

16 Hochhuth C, Berkovic D, Eibl H, Unger C, Doeneke D: Effects of antineoplastic phospholipids on parameters of cell differentiation in U937 cells. J Cancer Res Clin Oncol 1990;116:459–466.

17 Klein P, Hutter KJ, Berger MR, Schmähl D: *In vitro* sensitivity of different mammary carcinoma cell clones against hexadecylphosphocholine. J Cancer Res Clin Oncol 1990;116(Suppl I):618.

18 Berger MR, Schmähl D, Unger C, Eibl H: Antineoplastic activity of alkylphosphocholine derivatives in methylnitrosourea-induced rat mammary carcinoma. Proc Am Assoc Cancer Res 1988;29:330.

19 Scherf HR, Schuler J, Berger MR, Schmähl D: Therapeutic activity of ET-18-OCH_3 and hexadecylphosphocholine against mammary tumors in BD-VI rats. Lipids 1987;22:927–929.

20 Muschiol C, Berger MR, Schuler B, Scherf HR, Garzon FT, Zeller WJ, Unger C, Eibl H, Schmähl D: Alkylphosphocholines: Toxicity and anticancer properties. Lipids 1987;22:930–934.

21 Hilgard P, Stekar J, Voegeli R, Engel J, Schumacher W, Eibl H, Unger C, Berger MR: Characterisation of the antitumor activity of hexadecylphosphocholine (D 18506). Eur J Cancer Clin Oncol 1988;24:1457–1461.

22 Engel J, Schumacher W, Hilgard P, Stekar J, Penkert M, Schmähl D, Berger MR, Eibl H, Unger C, Nagel G: Hexadecylphosphocholine. Drugs Fut 1988;13:948–951.

23 Hilgard P, Stekar J, Voegeli R, Harlman JH: Experimental therapeutic studies with Miltefosine in rats and mice. Prog Exp Tumor Res. Basel, Karger, 1992, vol. 34, pp 95–109.

M.R. Berger, Institute of Toxicology and Chemotherapy,
German Cancer Research Center, Im Neuenheimer Feld 280,
D-W–6900 Heidelberg (FRG)

Eibl H, Hilgard P, Unger C (eds): Alkylphosphocholines: New Drugs in Cancer Therapy. Prog Exp Tumor Res. Basel, Karger, 1992, vol 34, pp 116–130

Experimental Therapeutic Studies with Miltefosine in Rats and Mice

Peter Hilgard, Jurij Stekar, Rainer Voegeli, Johannes H. Harleman

ASTA Pharma AG, Department of Experimental Cancer Research and Institute of Toxicology, Bielefeld, FRG

Introduction

Miltefosine (hexadecylphosphocholine) is the prototype of a new group of potential anticancer agents which belong to the chemical class of alkylphosphocholines. Its development was the logical consequence of investigations into the biochemistry of alkyllysophospholipids [1]. These latter compounds were shown to exert significant antitumor activity in appropriate experimental models [2]. Apart from direct cytotoxic effects on tumor cells, it was repeatedly suggested that these substances could also augment the immune response mounted by the host against the tumor [3]. Since the discovery that macrophage activation went along with the appearance of a phospholipase which could degrade phosphatidylcholine and phosphatidyletholamine to lysoderivatives and free fatty acids, alkyllysophospholipids were thought to be mediators of an immune response [4]. Based on the assumption that miltefosine's spectrum of activity might be similar to that of the alkyllysophospholipids, various pharmacological investigations were carried out in our laboratories during the last years. They are summarized in the subsequent communication.

In vitro Activity

In order to characterize the in vitro spectrum of activity of miltefosine against a selection of established rodent and human tumor cell lines, the compound was tested in conventional soft agar stem cell assays with exposure times between 7 and 12 days. The concentrations of miltefosine which inhibited colony formation of various cell lines by 90% (IC_{90}) are given in table 1. As can be seen the human KB and HL-60 cell line as well

Table 1. In vitro activity of miltefosine in the clonogenic assay using human and rodent tumor cell lines

Clonogenic assay	IC_{90} $\mu g/ml$
L1210 mouse leukemia	3.16
NK-mouse lymphoma	> 100
Ah 13s rat hepatoma	10
KB human carcinoma	0.7
HL60 human leukemia	2.8
CO 115 human colonic carcinoma	10
Mouse bone marrow	> 100

Table 2. Summary of in vivo activity of miltefosine in various transplantable tumors

Tumor	Site	Animal	Result
DS carcinosarcoma	s.c.	rat	negative
Leukemia L5222	i.p.	rat	negative
Leukemia L1210	i.p.	mouse	negative
Leukemia P388	i.p.	mouse	negative
B-16 melanoma	s.c.	mouse	negative
Lewis lung carcinoma	s.c.	mouse	negative
Lewis lung carcinoma	i.v.	mouse	marginal
MOP-C plasmocytoma	s.c.	mouse	marginal
KB (human)	s.c.	nu/nu mouse	active

as the mouse leukemia L1210 were amongst the more sensitive cell lines whereas the NK mouse lymphoma and normal mouse bone marrow were completely resistant to the cytotoxic action of miltefosine. A rat hepatoma and a human colon carcinoma line were moderately sensitive.

Screening Tumors in vivo

In contrast to the in vitro data it turned out that none of the conventional rodent screening tumors were sensitive to miltefosine treatment in vivo (table 2). Marginal activity was seen in intravenously transplanted Lewis lung carcinoma and in the solid MOP-C plasmocytoma. In view of the in vitro activity, the lack of in vivo effects in the L1210 leukemia is particularly interesting. Dose escalations of miltefosine to sublethal levels were performed in this model, yet there was no hint of antitumor activity. According to the pharmacokinetic studies reported

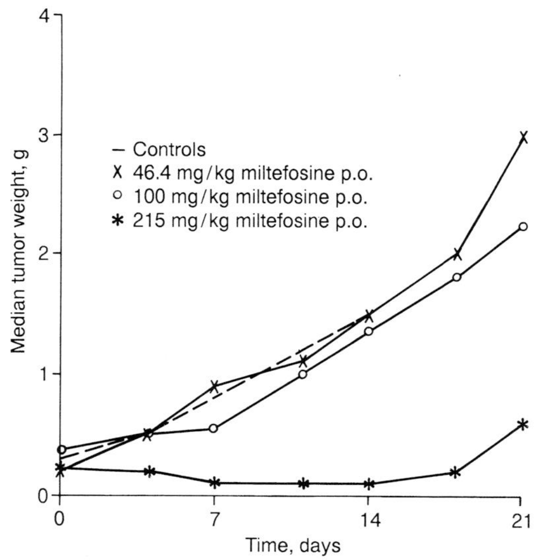

Fig. 1. Antineoplastic effect of various single oral miltefosine doses (administered at t = 0) on transplanted KB tumors of the nude mouse.

elsewhere in this volume, the animals' plasma levels should have reached and maintained cytotoxic concentrations. The discrepancy between in vitro and in vivo effects in the L1210 mouse leukemia cannot be explained at present.

KB Tumors in Nude Mice

In contrast to the findings in the L1210 system there was a very good correlation between in vitro and in vivo sensitivity of the KB tumor. This is an established human squamous cell carcinoma cell line which can be grown as solid tumor in NMRI nu/nu mice. The tumor growth curves following a single oral administration of various miltefosine doses are demonstrated in figure 1. With this dose schedule the dose efficacy relationship seems to be rather steep since the two lower doses did not differ from the controls whereas the highest dose completely inhibited the tumor cell growth. As can be seen from figure 2, two treatments with miltefosine (215 mg/kg) spaced 7 days apart resulted in a reduction of large established KB tumors by almost 90%. This sensitivity of the KB tumor is the most significant since in our experience this transplantable tumor shows a relatively poor sensitivity to cyclophosphamide and cisplatin.

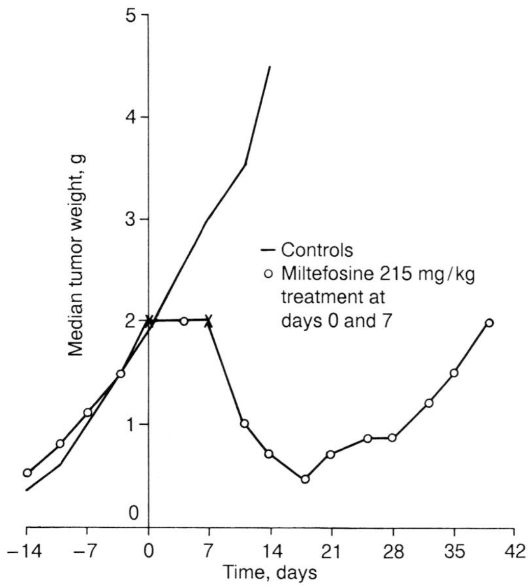

Fig. 2. Antineoplastic action of 2 oral miltefosine doses (215 mg/kg) on s.c. KB tumors in nu/nu NMRI mice.

The experiments described so far indicate that the mode of action of miltefosine is distinctly different from that of conventional cytostatic agents. The enormous differences in sensitivity of different cell lines to the compound as well as the unusual discrepancy between in vitro and in vivo activity indicate that more complex mechanisms might be associated with the antitumor activity of miltefosine.

Immunological Investigations

If immunological mechanisms played a major role in the miltefosine-mediated antitumour response, they would have to be more or less independent of the T cell system because, as described above, the compound was highly active in thymus aplastic nude mice bearing KB tumors. Further studies into specific aspects of the immune system of rats and mice did not reveal any major activity of miltefosine in the experimental models used. If spleen cells from miltefosine-treated rats were employed as effector cells there was no increased NK cell activity as measured by the chromium release of YAC-1 cells, the classical target for natural killer activity.

Likewise, humoral antibody production following the immunization with sheep red blood cells was not changed by miltefosine pretreatment. A possible action on macrophages in vitro was investigated by the phagocytic activity of mouse bone marrow macrophages. Again, no stimulatory activity was seen. In conclusion, to date there is no experimental evidence that miltefosine's antitumor activity is mediated by the host's immune system [5].

Autochthonous Rat Tumors

The data of Muschiol et al. [6] from the German Cancer Research Center clearly showed that miltefosine and other alkylphosphocholines exhibited substantial anticancer activity in vivo in the methylnitrosourea (MNU)-induced rat mammary carcinoma. These findings suggested that miltefosine might also be active in the classical, hormone-dependent dimethylbenz(a)anthracene (DMBA)-induced rat mammary carcinoma. This autochthonous tumor was induced by application of 20 mg DMBA to 50-day-old female Sprague-Dawley rats. If treatment was started when the tumor weight had reached approximately 0.2 g and miltefosine was administered twice weekly for altogether 2 weeks the suppression of tumor growth turned out to be strictly dose dependent (fig. 3). Similar to the KB tumor in nu/nu mice large established DMBA tumors could also be brought into remissions by oral miltefosine treatment (fig. 4).

If miltefosine was given in a 'prophylactic intent' directly following the administration of DMBA, the appearance of tumors could be delayed as long as the animals were treated (fig. 5). On average, 50 days after cessation of therapy the tumors appeared with the same distribution and growth kinetics as in previously untreated controls. This observation indicated that miltefosine effectively inhibited cell proliferation and did not interfere with the carcinogenic action of DMBA.

The knowledge about the schedule dependency of a compound's efficacy is of certain clinical importance and, therefore, this type of investigation was carried out in DMBA tumor-bearing rats. The weekly dose of 200 mg/kg body weight was given to the animals in various schedules and the respective tumor growth curves are presented in figure 6. As can be seen, the antitumor effect was completely independent of the schedule, a weekly single dose was as efficacious as the daily oral dosing. This finding led to the conclusion that the plasma half-life of the compound in rats must be relatively long. Subsequent pharmacokinetic studies have fully confirmed this assumption.

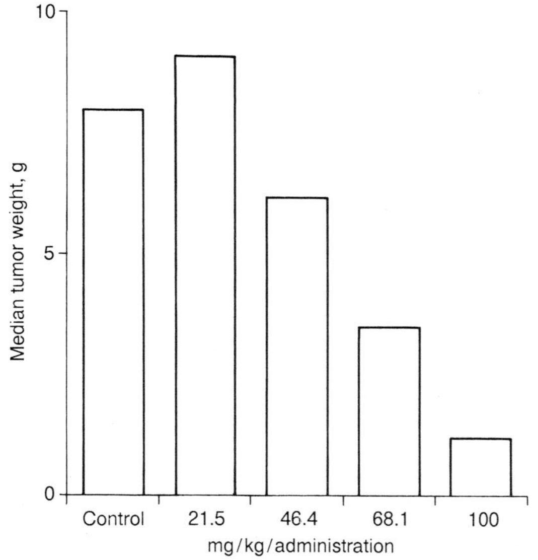

Fig. 3. Median tumor weights of DMBA-induced tumors on day 14 after oral Miltefosine treatment (days 1, 5, 9 and 13) in different doses.

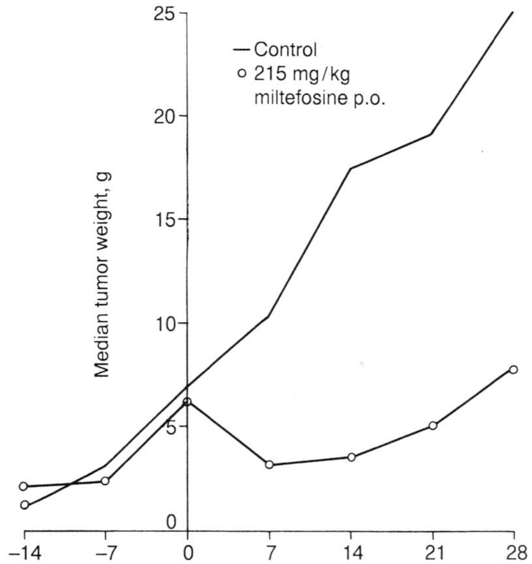

Fig. 4. Treatment of advanced DMBA tumors with a high single oral dose of miltefosine. Group size: 7 animals.

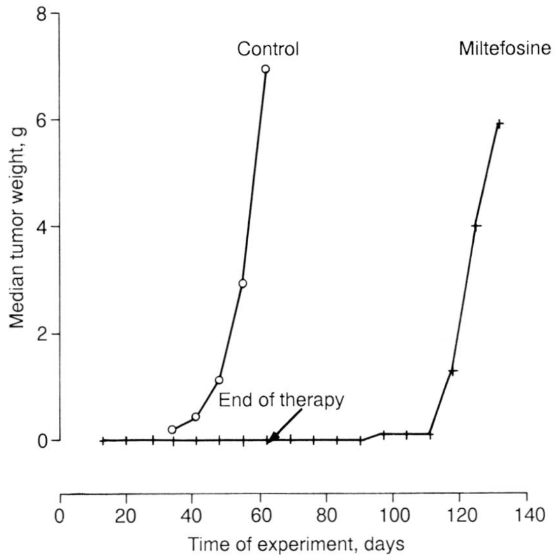

Fig. 5. 'Prophylactic' treatment with oral miltefosine (200 mg/kg/1 × week) in DMBA-induced rat mammary carcinomas (n = 10/group).

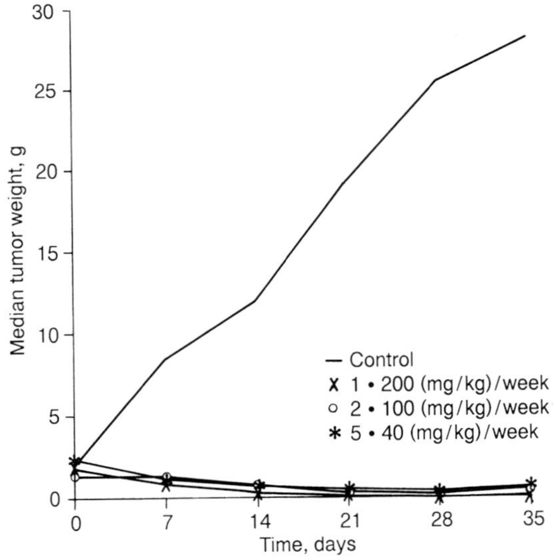

Fig. 6. Effect of different dose schedules of oral miltefosine on the response of DMBA-induced rat mammary carcinomas.

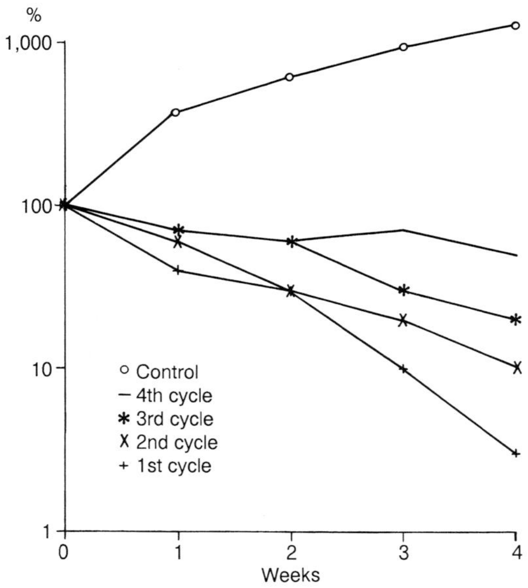

Fig. 7. Growth curves of DMBA-induced rat mammary carcinomas after repeated cycles of miltefosine (200 mg/kg/week × 4).

Miltefosine Resistance

So far, attempts in our laboratory to produce cell lines resistant to miltefosine by continuous in vitro exposure of the compound have failed. We therefore decided to investigate the potential development of drug resistance in vivo. DMBA tumor-bearing animals were treated for 4 weeks with a weekly 200 mg/kg dose schedule. All animals went into complete remission or had a prolonged period of stationary tumor growth. When the tumors started to regrow they were treated again with the same schedule for a second time. This procedure was altogether repeated 4 times. Figure 7 shows the average tumor growth curves during the different treatment cycles. Obviously, the therapy became less effective after intensive pretreatment, but, the classical picture of drug resistance was never observed. The reason for the gradual loss of the tumor's sensitivity to miltefosine after repeated treatments remains to be elucidated.

Morphological Effects on DMBA Tumors

Throughout a 3 week study of daily oral treatment with 20 mg/kg miltefosine serial histologies of DMBA-induced mammary tumors were

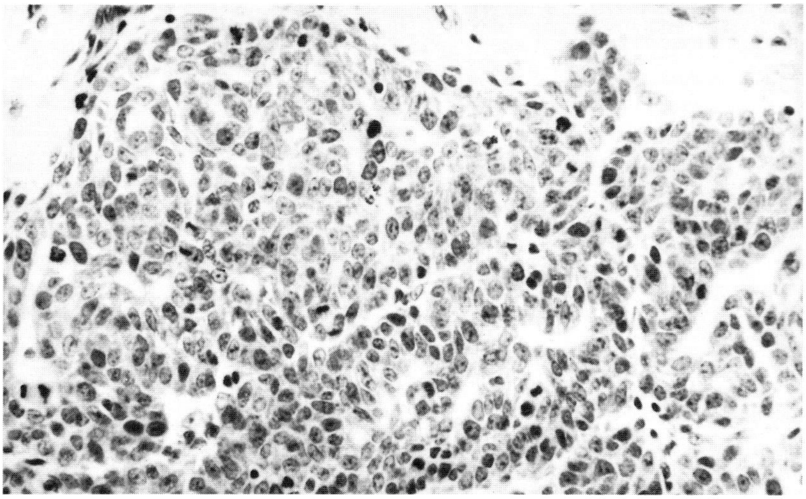

Fig. 8. Histological section of an untreated DMBA-induced rat mammary carcinoma. Explanation see text.

carried out as long as there was a palpable tumor present. Tumors of untreated control animals showed the typical features of a tubular adenocarcinoma with a high mitosis rate and invasive growth at the periphery (fig. 8). Already 3 days after initiation of miltefosine therapy the mitotic rate decreased and first signs of epithelial differentiation became evident. Finally, mitosis disappeared almost completely in the tumors of treated animals and the tubuli had a well-defined, lumen-orientated epithelial layer with a tendency to form a single cell lining.

The reduction in tumor mass was invariably associated with the formation of subepithelial nests of fusiform epithelial cells which separated themselves from the luminal orientated epithelial cells. Both within the luminal epithelium and the subepithelium nests of apoptotic cell bodies were frequently seen. Figure 9 shows a typical DMBA tumor after 14 days treatment. Figure 10 shows a DMBA tumor after 28 days of treatment. The remaining tumor mass was well demarcated from the surrounding tissue and within the tubular epithelium secretory material was frequently seen. The relative amount of tumor stroma increased gradually throughout treatment. The histological changes observed were compatible with a gradual differentiation of the tumor towards normal secreting mammary gland tissue. This in vivo observation fully confirms previous in vitro data which have shown that miltefosine is able to induce cellular differentiation

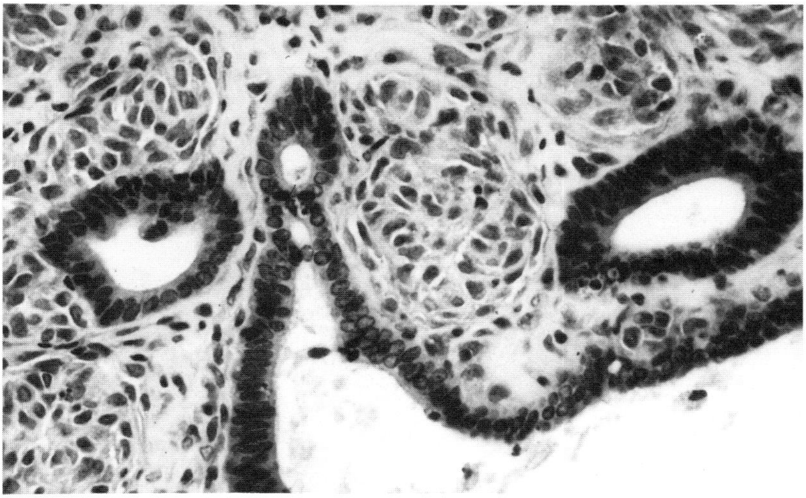

Fig. 9. Histological section of a DMBA-induced rat mammary carcinoma after 14 days of oral miltefosine treatment.

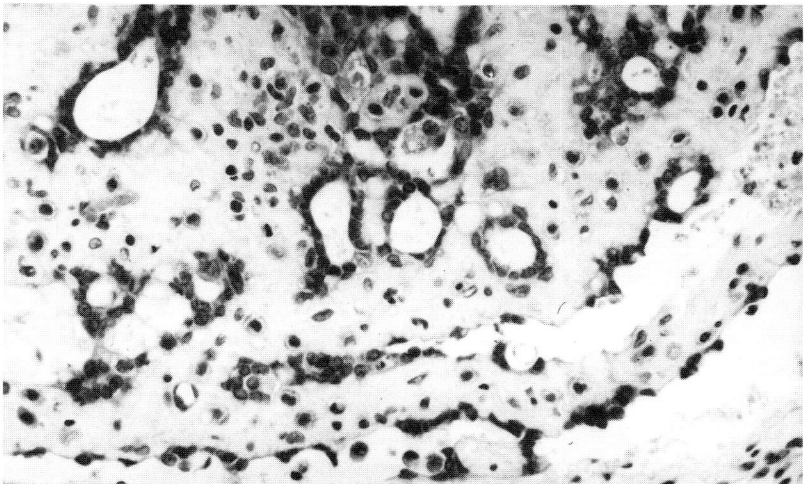

Fig. 10. Histological section of a DMBA-induced rat mammary carcinoma after 28 days of oral miltefosine treatment. Explanation see text.

in tumor cells with appropriate markers [7]. It is conceivable that the induction of tumor differentiation towards normal tissue constitutes one of the mechanisms by which miltefosine brings about tumor regressions.

Hormone Insensitive DMBA Tumors

Since both the DMBA- as well as the MNU-induced mammary carcinomas are well-established models of hormone-dependent tumors, the possibility that miltefosine's action was mediated by endocrine effects had to be ruled out. The most effective treatment of DMBA tumors is ovariectomy [8]. However, most of the animals will relapse after 3–6 months. These second tumors are usually hormone receptor negative and do not respond to hormonal therapy. This is shown in figure 11 where it can be seen that tamoxifene was completely inactive in DMBA tumors relapsed after ovariectomy. In contrast, these tumors still responded to miltefosine, indicating that its therapeutic activity is largely independent from the hormone receptor status of the tumor.

Lymphokine Activation?

The toxicology of miltefosine is well characterized and the doses and schedules used in the pharmacological experiments described here were generally well tolerated (ASTA Pharma AG, internal reports). Nevertheless, cumulative doses could induce loss of appetite ultimately leading to complete refusal of food uptake by the animals. An intriguing phenomenon of possible clinical implications was the fact that the tolerance of miltefosine was strictly dependent on the presence or absence of a tumor as well as on the overall size of the tumor mass. Figure 12 shows the body weight change of tumor-bearing and normal Sprague-Dawley rats with and without miltefosine treatment. The most dramatic change in body weight occurred in animals with large tumors. The significance of this observation remains speculative but could indicate that miltefosine-induced weight loss was mediated by tumor necrosis factor (cachectin)-like substances. An additional hint that lymphokines might be involved in the pharmacological and toxicological properties of miltefosine came from the observation of a granulocytosis during miltefosine treatment which strongly resembled the known effect of G-GSF (granulocyte colony-stimulating factor).

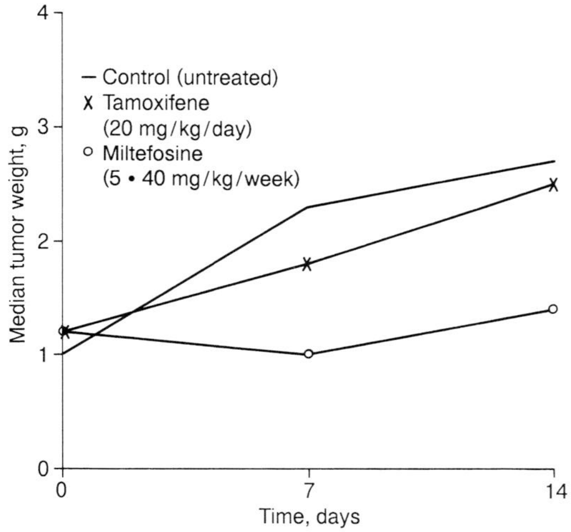

Fig. 11. Treatment of DMBA-induced rat mammary carcinomas relapsed after ovariectomy with tamoxifene and miltefosine (n = 7/group).

Topical Miltefosine Treatment

A useful clinical application of miltefosine is its topical administration on cutaneous breast cancer metastases. Although this therapy has no curative intent it may be of considerable palliative value to the patient. In order to be able to investigate the various pharmacological and toxicological effects of a topical treatment, an in vivo model with subcutaneously transplanted KB cells in nude mice was established. After the tumor had reached a size of approximately 0.1–0.15 g, the animals were randomized and treated.

Miltefosine was formulated in a mixture of alkylglycerols according to Unger and Eibl (this volume) and applied using an epicutaneous patch which was placed over the tumor. The patch was left for 6 h and the treatment was repeated daily for 5 days. Local treatment with the above miltefosine formulation was found to be effective. Eleven days after initiation of the therapy, the median tumor weight in the treatment group was only 30% of that of the control. Contralateral miltefosine administration ruled out that the observed effects were brought about by systemic absorption. The model appeared to be a suitable test for topical tumor treatment and was able to experimentally establish the effect of epicutaneous miltefosine application.

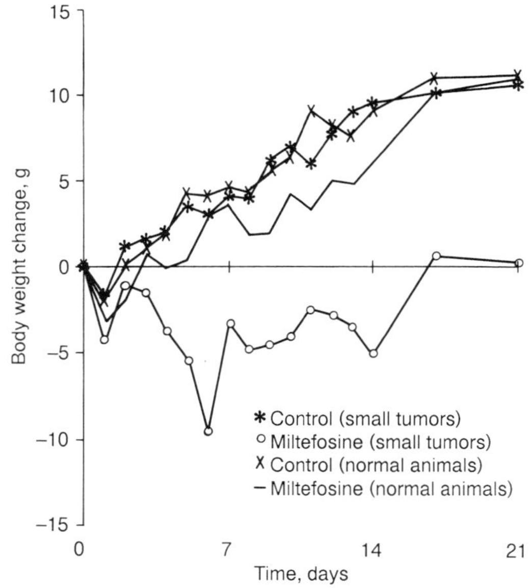

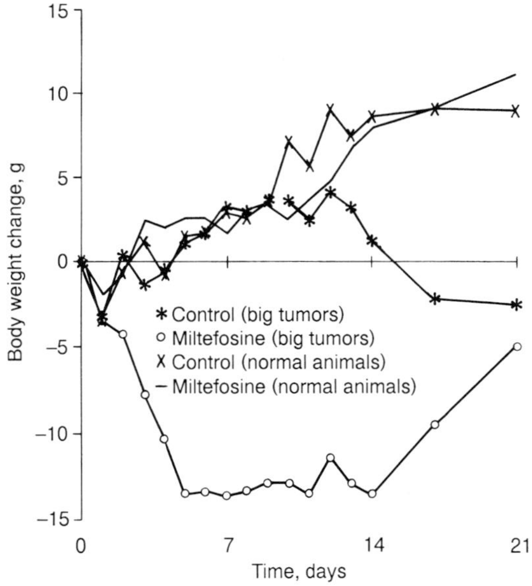

Fig. 12. Body weight change with and without miltefosine (14.7 mg/kg/day × 14) in normal as well as in small or large DMBA tumor-bearing Sprague-Dawley rats.

Conclusions

The experimental profile of miltefosine which was described above clearly indicates that the mechanism by which miltefosine exerts its antitumor activity must be distinctly different from that of other anticancer agents. It also seems that there are substantial differences between various alkyllysophospholipids and miltefosine. Octadecylmethyl-rac-glycerophosphocholine (ET_{18}-OCH_3) was the first alkyllysophospholipid to undergo extensive preclinical testing. It was repeatedly shown to exhibit immunomodulatory activity, a property which was considered to be of importance for the antitumor activity of this compound in vivo [9]. This is in sharp contrast to miltefosine which showed no such activity in appropriate test systems. Furthermore, the pronounced selectivity of miltefosine for certain experimental tumors is outstanding. Neither ET_{18}-OCH_3 nor its thioanalog, ilmofosine, exhibit such a clear-cut antitumor spectrum [10]. Whereas the latter agents showed some activity in transplantable rodent tumors such as P388, Lewis lung and L1210, the miltefosine action was confined to models like autochthonous rat tumors and certain xenografts like the KB tumor. This might indicate that there is a more specific molecular target for miltefosine in comparison to the other lysophospholipid derivatives.

In vitro studies with NIH-3T3 cell system revealed that miltefosine inhibited protein kinase C activity in cell free extracts as well as in intact cells [11]. The mechanism of inhibition was a competition with the biological activator phosphatidylserine. In addition, there is evidence that miltefosine acts as an inhibitor of phosphoinositidase C. Taking these pieces of evidence together, it becomes likely that the antiproliferative effect of miltefosine is connected to an interference with the polyphosphoinositide metabolism mediated mitogenic signal transduction. This would at least partially explain why cell systems which grow independently from mitogenic stimuli like P388, L1210 and Lewis lung carcinoma were insensitive to miltefosine in vivo.

References

1 Eibl H, Unger C: Phospholipide als Antitumormittel: Möglichkeiten einer selektiven Therapie. Aktuel Onkol 1987;34:1–18.

2 Berdel WE, Bausert WRE, Fink V, Rastetter J, Munder PG: Antitumor action of alkyllysophospholipids (review). Anticancer Res 1981;1:345–352.

3 Andreesen R, Osterholz J, Luckenbach GA, Costabel U, Schulz A, Speth V, Munder PG, Löhr GW: Tumor cytotoxicity of human macrophages after incubation with synthetic analogues of 2-lysophosphatidylcholine. J Natl Cancer Inst 1984;72:53–59.

4 Munder PG, Fischer H, Weltzien HV, Oettgen HF, Westphal O: Lysolecithin analogs: A new class of immunopotentiators with antitumor activity. Proc Am Assoc Cancer Res 1976;17:174.

5 Hilgard P, Kampherm E, Nolan L, Pohl J, Reissmann T: Investigations into the immunological effects of miltefosine a new anticancer agent under development, J Cancer Res Clin Oncol, in press.

6 Muschiol C, Berger MR, Schuler B, Scherf HR, Garzon FT, Zeller WJ, Unger C, Eibl HJ, Schmähl D: Alkyl phosphocholines: Toxicity and anticancer properties. Lipids 1987;22:930–934.

7 Hilgard P, Harleman JH, Voegeli R, Maurer HR, Echarti C, Unger C: The antineoplastic activity of hexadecylphosphocholine (HPC) is associated with tumor cell differentiation. Proc Am Assoc Cancer Res 1989;30:580, abstr No 2310.

8 Leclercq G, Heuson JC: Specific estrogen receptor of the DMBA-induced mammary carcinoma of the rat and its estrogen-requiring molecular transformation. Eur J Cancer 1973;9:675–680.

9 Andreesen R, Giese V: Differential effects of ether lipids on the activity and secretion of interleukin-1 and interleukin-2. Lipids 1987;22:836–841.

10 Pahlke W: Präklinische Untersuchungen des neuen Thioätherphospholipids BM 41.440. Aktuel Onkol 1987;34:65–76.

11 Überall F, Oberhuber H, Maly K, Zaknun J, Demuth L, Grunicke HH: Hexadecylphosphocholine inhibits inositol phosphate formation and protein kinase C activity. Cancer Res 1991;51:807–812.

Peter Hilgard, ASTA Pharma AG, Head, Cancer Res, PO Box 14 01 29,
D-W–4800 Bielefeld 14 (FRG)

Eibl H, Hilgard P, Unger C (eds): Alkylphosphocholines: New Drugs in Cancer Therapy. Prog Exp Tumor Res. Basel, Karger, 1992, vol 34, pp 131–142

Hexadecylphosphocholine and Octadecyl-methyl-glycero-3-phosphocholine: A Comparison of Hemolytic Activity, Serum Binding and Tissue Distribution

J. Kötting[a], *N.W. Marschner*[b], *W. Neumüller*[a], *C. Unger*[b], *H. Eibl*[a]

[a]Max-Planck-Institut für biophysikalische Chemie, and [b]Medizinische Universitätsklinik, Abteilung für Hämatologie/Onkologie, Göttingen FRG

Introduction

The successful application of hexadecylphosphocholine (HePC) in the topical treatment of skin metastases of mammary carcinomas [1, 2] in phase I/II trials was decisive to develop HePC as an antitumour drug. Starting from alkylglycerophosphocholines Eibl and Unger [3] searched for the minimal structural requirements for antineoplastic activity, and have demonstrated [4] that glycerol is not essential for this effect. The final decision for the development of HePC was based on enzymatic in vitro assays [5, 6], cell culture studies on leukemic cell lines [7] and animal experiments with methylnitrosourea-induced autochthonous rat mammary carcinoma [8]. As a prerequisite for the further application of HePC and homologues as drugs against malignant diseases, a pharmacological and toxicological evaluation is necessary. This is especially true in the perspective of a systemic administration of alkylphosphocholines. In this publication, we present the first results for HePC in comparison to 1-O-octadecyl-2-O-methyl-*sn*-glycero-3-phosphocholine.

Material and Methods

Chemicals

Sodium dodecylsulfate (SDS) was purchased from SIGMA Chemicals (Munich, FRG). Human serum albumin (purified) was obtained from Behring Werke (Marburg, FRG). All other chemicals were products of Merck (Darmstadt, FRG). The solvents were obtained from Baker Chemicals (Deventer, NL) and used without further purification.

Synthesis of Alkylphospholipids

The lipids 1-palmitoyl-*sn*-glycero-3-phosphocholine (lysolecithin, LPC), $ET_{18}OCH_3$ and 1-O-hexadecyl-2-acetyl-*sn*-glycero-3-phosphocholine (platelet-activating factor, PAF) were synthesized in our laboratory according to methods described earlier [9, 10]. The synthesis of HePC is described by Eibl and Engel [this volume]. The purity of the compounds was better than 99% as demonstrated by thin-layer chromatography.

Hemolysis Experiments

The procedure of the hemolysis assay is described in detail elsewhere [11, 12]. Erythrocytes from fresh human blood were separated from plasma by centrifugation (800*g*, 10 min), then washed three times with 0.01*M* sodium phosphate buffer (pH 7.4), containing 0.9% NaCl (PBS) and diluted with PBS to a suspension containing $1.5 \cdot 10^7$ cells/ml. The cell count was performed in a Neubauer chamber.

The light absorption at 560 nm of a 1-ml erythrocyte suspension at 37 °C was continuously recorded in a single beam diode array spectrophotometer (Hewlett Packard 8411). The value of absorbance for the cell suspension in phosphate buffer in the absence of lysolipids was defined as A_0 and corresponds with 0% hemolysis; the numerical value is around 1.75 for $1.5 \cdot 10^7$ cells/ml in a 10 mm cuvette. A value of 0.3 was observed for total hemolysis after addition of 10 μl 10% SDS solution to 1 ml of the erythrocyte suspension. This absorbance value was defined as A_{100} and corresponds to 100% hemolysis. Thus, a difference in absorption values of 1.45 (A_0–A_{100}) describes 100% hemolysis in a suspension containing $1.5 \cdot 10^7$ cells/ml. Addition of lysolipids or alkylphosphocholines led to hemolysis which resulted in a decrease of the absorbance value. For all compounds tested it was observed that the absorbance reached a constant value after 2 min of incubation. The percentage hemolysis (H%) obtained with lyso compounds (A_L) at different concentrations can be calculated by using the following equation:

$$H\% = \frac{(A_0 - A_L)}{1.45} \cdot 100\%,$$

Preparation of Lipoproteins

Lipoproteins were prepared by density gradient ultracentrifugation from fresh human serum according to the method described by Armstrong et al. [13]. In a second set of experiments samples of fresh human serum were spiked with 100 nmol/ml HePC or $ET_{18}OCH_3$, respectively, before separation of the serum protein fractions.

Animal Experiments

For each compound and time point 3 rats (Wistar strain, female, 180–220 g) were fed by gastric intubation with 15 μmol (30 mg/kg) of HePC or $ET_{18}OCH_3$ over periods of 6 and 18 days. Schedule for drug administration was feeding 6 days a week, omitting Sunday. Animals were killed by CO_2 anesthesia 24 h after the last application of the drugs. The uptake of HePC and $ET_{18}OCH_3$ in sera and different organs of rats were analyzed by high-performance thin-layer chromatography. This method is described by Kötting et al. [this volume].

Results and Discussion

It is now well established that phospholipids are absorbed in the gastrointestinal tract of animals as lyso compounds [14–16]. The nutritional

Table 1. Critical micellar concentration (CMC) of various synthetic phospholipids and alkylphosphocholines: values were estimated with N-phenyl-1-naphthylamine according to Neumüller [21]

Compounds	CMC [10^{-6} *M*]
1-Palmitoyl-*sn*-glycero-3-PC	1.6
1-O-hexadecyl-*sn*-glycero-3-PC	1.3
1-Palmitoyl-propanediol (1,3)-PC	1.0
1-O-hexadecyl-propanediol (1,3)-PC	1.3
$ET_{18}OCH_3$	0.4
Dodecyl-PC	210
Tetradecyl-PC	25
HePC	8.0
Octadecyl-PC	3.4
Oleyl-PC	2.8

PC = Phosphocholines.

phospholipids are degraded by phospholipase A_2, phosphodiesterase and phosphatase producing free fatty acids, monoglycerides, glycerol and phospholipids [16]. Together with bile acids lysophospholipids solubilize other breakdown products as mixed micelles. Subsequently, after the uptake by mucosa cells, lipids are resynthesized, incorporated into lipoproteins, mainly chylomikrons and transported by the lymphatic system (ductus thoraticus) to the blood circuit [17]. Only glycerol and short chain fatty acids are transferred directly to the blood [16].

Absorption of HePC and $ET_{18}OCH_3$ is conceivable in the same manner as for natural lysophospholipids, since HePC and $ET_{18}OCH_3$ are forming micellar structures with critical micelle concentrations (CMC) comparable to those of lysolecithins (table 1). For instance, 1-palmitoyl-*sn*-glycero-3-phosphocholine, 1-O-hexadecyl-*sn*-glycero-3-phosphocholine, 1-palmitoyl-propanediol(1,3)-phosphocholine, and 1-O-hexadecyl-propanediol (1,3)-phosphocholine have very similar CMC values of about $2 \cdot 10^{-6}$ *M*. The CMC value of HePC is $8 \cdot 10^{-6}$ *M* and much higher than the values for the respective lysolecithins. This indicates that the C_3 unit of glycerol contributes to the apolarity of these molecules. To reach a CMC value near 10^{-6} *M* with alkylphosphocholines, a chain length of 20 carbon atoms is required as demonstrated for eicosanoylphosphocholine. Consequently, $ET_{18}OCH_3$ has a CMC value of $0.5 \cdot 10^{-6}$ *M* since in this molecule the apolarity of the C_{18} chain is increased by additional carbon atoms of the glycerol unit. Roughly, the hemolytic concentrations of lysophospholipids in serum-free assays coincide with the CMC.

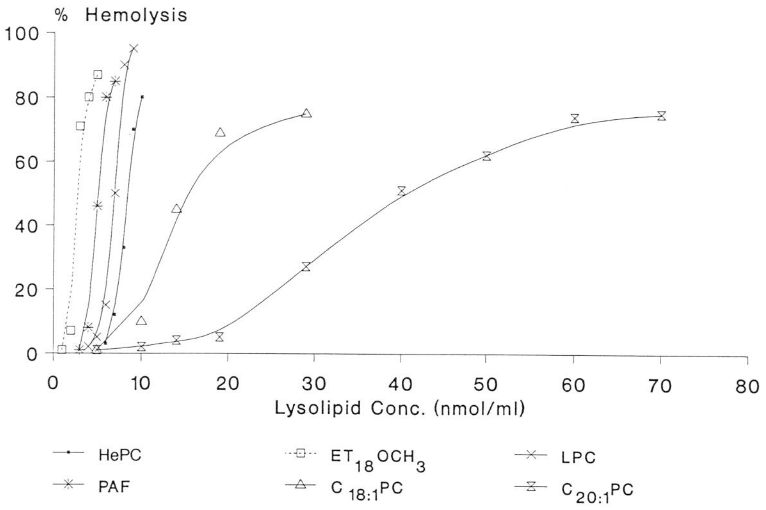

Fig. 1. Spontaneous hemolysis of $1.5 \cdot 10^7$ erythrocytes/ml in PBS at 37 °C caused by different lysolipids and alkylphosphocholines within 2 min. Hemolysis was followed by measuring the decrease of absorbance at 546 nm. The 100% hemolysis value was estimated by adding 10% sodium dodecylsulfate to the erythrocyte suspension. PC = phosphocholine; LPC = lysolecithin; $C_{18:1}$ = oleyl; $C_{20:1}$ = eicosanoyl.

Human eythrocytes ($1.5 \cdot 10^7$ cells/ml) in PBS at 37 °C were incubated with various amounts of different APC and lysophospholipids and hemolysis was recorded after 2 min. The red blood cell concentration in the test cuvette was 300 times lower than in human blood. Figure 1 depicts the hemolysis curves induced by different lysolipids and alkylphosphocholines. In a serum-free system, 50% hemolysis is observed for concentrations of $2 \cdot 10^{-6}$ *M* for $ET_{18}OCH_3$, $4 \cdot 10^{-6}$ *M* for PAF, $7 \cdot 10^{-6}$ *M* for 1-palmitoyl-lysolecithin and $8 \cdot 10^{-6}$ *M* for HePC. Unsaturated longer chain alkylphosphocholines as oleylphosphocholine (15 μM) and eicosanoyl phosphocholine are significantly less hemolytic (40 μM) in this assay.

The hemolysis induced by different lysolipids and alkylphosphocholines was inhibited by addition of serum to the assay mixture. This is shown in figure 2a for HePC- and 2b for $ET_{18}OCH_3$-induced hemolysis. Addition of 10 μl serum increased the HePC concentration necessary for 50% hemolysis from 8 up to 60 nmol/ml (fig. 2a). Comparing the slopes of the total serum curves in figures 2a and 2b, it is obvious that HePC- as well as $ET_{18}OCH_3$-induced hemolysis was inhibited to the same extent. However, the inhibitory potency of the major serum proteins is different for

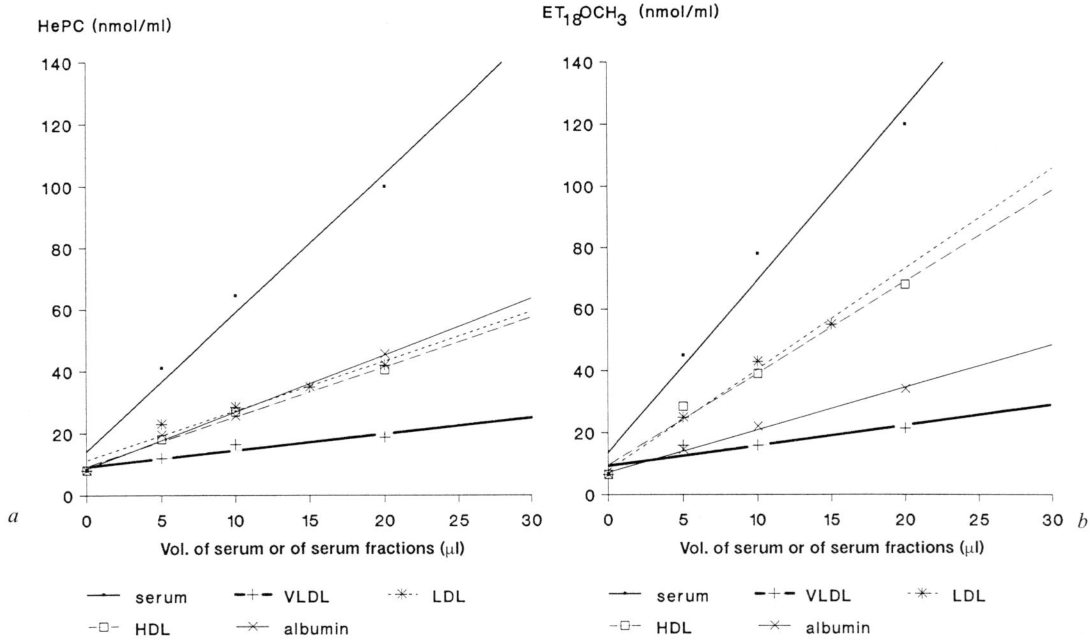

Fig. 2. Influence of human serum albumin and serum protein fractions on the concentration of HePC (*a*) and $ET_{18}OCH_3$ (*b*) necessary to induce 50% hemolysis. Hemolysis assays were performed as described in figure 1. The indicated amounts of serum or serum fractions were added to the erythrocyte suspension shifting hemolysis curves to higher lipid concentrations.

HePC and $ET_{18}OCH_3$. The contribution of the individual lipoproteins and albumin to the serum binding of HePC and $ET_{18}OCH_3$ was studied by separation of the major lipoprotein fractions by differential ultracentrifugation of human serum according to Armstrong et al. [13]. The separated lipoprotein subfractions were adjusted to their original concentration in serum with PBS to make sure that x-axis scales for each serum protein in figure 2 are identical. While very-low-density lipoproteins (VLDL) had only minor influence on the hemolytic activity of HePC and $ET_{18}OCH_3$, binding to albumin played a more pronounced role in the case of HePC whereas low-density lipoprotein (LDL) and high-density lipoprotein (HDL) fractions of the serum were more important in inhibiting $ET_{18}OCH_3$-induced hemolysis.

With the experimental set up to measure the inhibition of lysolipid-induced hemolysis, binding capacities of albumin for different lysolipids were estimated. Figure 3 depicts the influence of purified human albumin

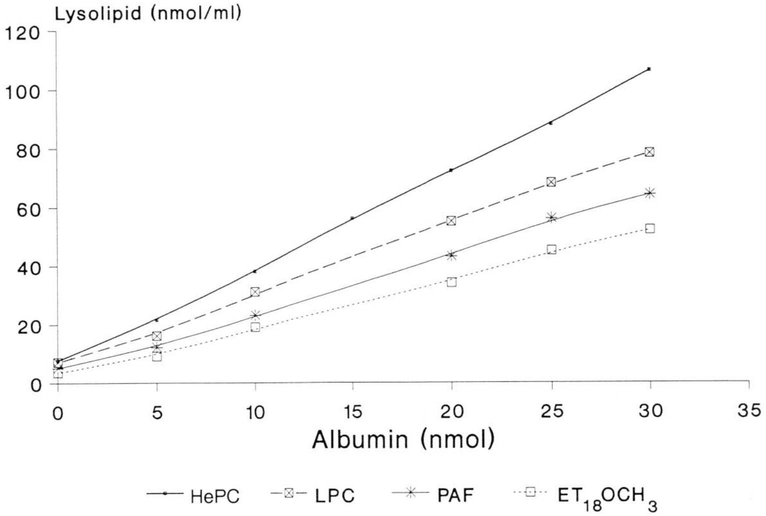

Fig. 3. Influence of purified human serum albumin on the alkyl-lysolipid and alkyl-phosphocholine concentrations inducing 50% hemolysis in the assay described in figure 1.

Table 2. Binding sites of serum albumin for different alkyl lysolipids and HePC estimated from figure 3, where the slope represents the numbers of molecules which at least can be bound by one molecule of serum.

Lysolipid	HePC	LPC	PAF	$Et_{18}OCH_3$
Slope	3.3	2.8	2.0	2.0

on the concentrations of lysolipids necessary to induce 50% hemolysis. HePC hemolysis is clearly more influenced by the addition of albumin than that of $ET_{18}OCH_3$. From the slopes of the different curves in figure 3, the binding sites of albumin for the respective lysolipids under the chosen conditions were calculated and summarized in table 2. Obviously, 1 mol albumin can bind 3.3 mol HePC, 2.8 mol 1-palmitoyl-*sn*-glycero-3-phosphocholine, 2.0 mol $ET_{18}OCH_3$ or 2.0 mol PAF. The similar binding behavior of $ET_{18}OCH_3$ and PAF most probably reflect close similarities in the chemical structure of these molecules. An explanation for the stronger binding of HePC to albumin in comparison to glycerol-containing lysolipids might be the structural relationship of HePC to fatty acids. One mol albumin can bind 4 moles fatty acid [18].

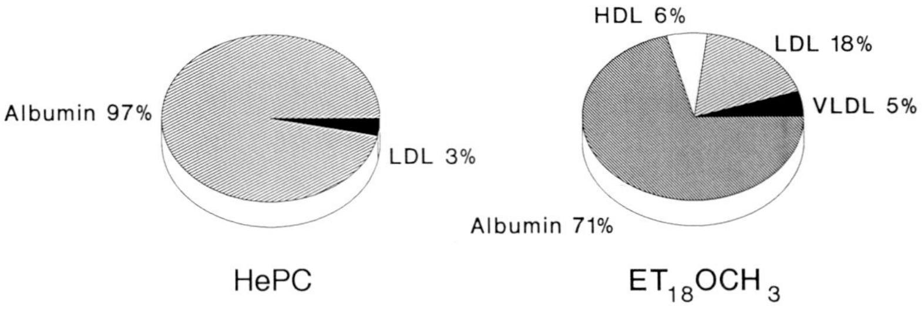

Fig. 4. Samples from fresh human serum were spiked with 100 nmol/ml of HePC or $ET_{18}OCH_3$, respectively. After separation of the serum protein fractions by density gradient ultracentrifugation [18], the amount of HePC and $ET_{18}OCH_3$ attached to the distinct protein fractions was measured by thin-layer chromatography [13]. In addition, the identity of the lipoprotein fractions was confirmed by analyzing their lipid composition by HPTLC.

Table 3. Absorption of HePC and $ET_{18}OCH_3$ in rats.

	HePC, μmol	$ET_{18}OCH_3$, μmol
Oral dose	5	5
Feces	0.05 (1%)	0.2 (4%)
Urine	< 1%	< 1%

For each compound, 3 Wistar rats were fed by gastric intubation with 5 μmol of the respective drug. Within 24 h all feces and urine excreted by the animals were collected and the content of HePC and $ET_{18}OCH_3$ was estimated after lipid extraction by HPTLC.

In addition, binding of HePC and $ET_{18}OCH_3$ to serum components was studied by incubation of 1 ml human serum with 100 nmol of the respective lysolipid. After 15 min, the serum proteins were separated by differential ultracentrifugation (fig. 4). HPTLC lipid analysis of the serum subfractions [13] showed that HePC was almost completely bound to serum albumin. In contrast, only 71% of $ET_{18}OCH_3$ were bound to the albumin fraction under these conditions. The other serum fractions contained substantial amounts of $ET_{18}OCH_3$. For instance 18% were incorporated in the LDL fraction, 6% in the HDL fraction and 5% in the VLDL fraction. This distribution is in coincidence with the hemolysis experiments shown in figure 2. The differences in the binding of HePC and $ET_{18}OCH_3$ to serum components may influence the serum levels and also the uptake of the compounds by different tissues.

The absorption efficiency in rats after oral application of 5 μmol HePC and 5 μmol $ET_{18}OCH_3$ is shown in table 3. The analysis of the feces

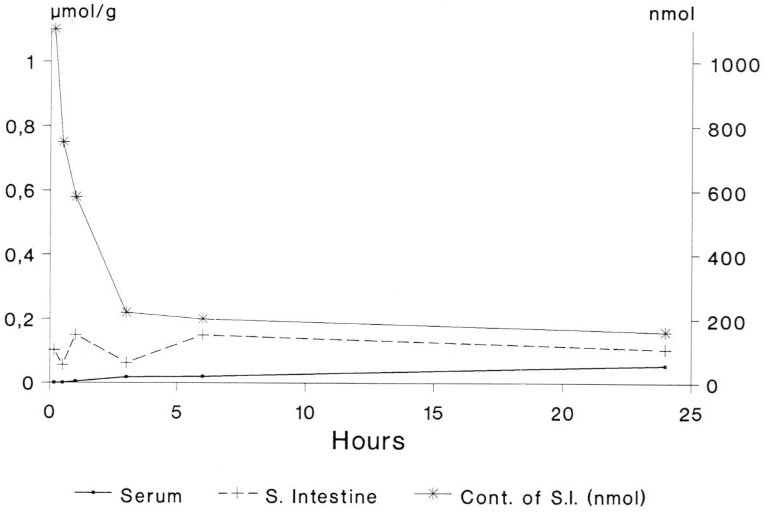

Fig. 5. Time course of HePC absorption in the small intestine. Wistar rats (180–220 g) were fed with 10 μmol HePC by gastric intubation. Starting 10 min after application, for each time point indicated, 2 animals were killed by CO_2 anesthesia, unbled and disected. The complete content of the small intestine (S.I.) was collected as well as pieces of duodenum, jejunum and ileum. The total amount of HePC in the small intestine content and the HePC concentration in serum and small intestine tissue samples were analyzed by thin-layer chromatography. The values are means of three estimations for each animal. These experiments were performed by J. Dreos (University Hospital, Göttingen, FRG).

collected within 24 h after feeding revealed that about 98% of HePC and 96% of $ET_{18}OCH_3$ were absorbed by the animals. Excretion in urine was negligible. After oral application of 10 μmol HePC, its concentration in the lumen of small intestine dropped within 3 h from 1,200 to 200 nmol (fig. 5). This was caused by absorption and not by intestinal metabolism, since incubations of HePC with tissue and contents of small intestine did not result in the formation of degradation products. In small intestine tissue, HePC concentration amounted to 100 nmol/g within the first 10 min after application and was almost constant for about 24 h. In serum, HePC was first detected 6 h after oral application. The concentration was 40 nmol/ml, and increased to 55 nmol/ml within 24 h. The time delay of 6 h supports the view that HePC transfer from the small intestine to serum occurs via the lymphatic fluid.

As shown in figure 6, after daily application of 15 μmol each of HePC (30 mg/kg) or $ET_{18}OCH_3$ (38 mg/kg), the serum concentration of HePC increased to 170 nmol/ml after 1 week, which may already indicate a steady

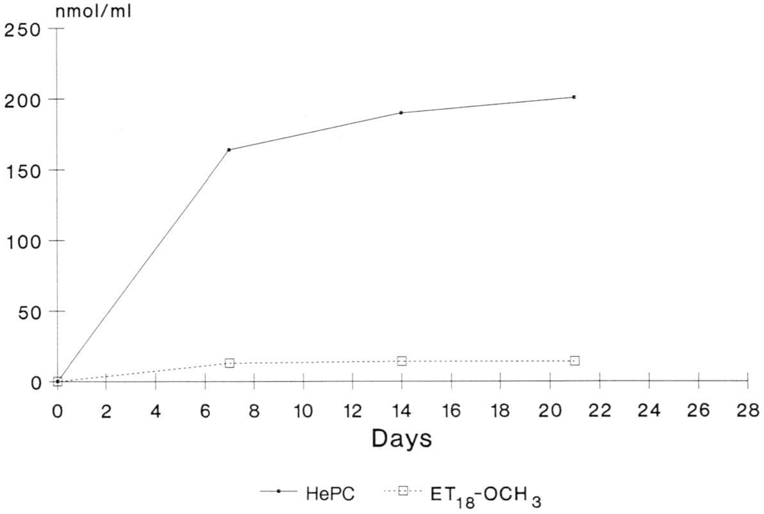

Fig. 6. Serum concentrations of HePC and $ET_{18}OCH_3$ in Wistar rats by daily p.o. application of 75 μmol/kg of the respective compound. SD was ± 10%.

state situation. For $ET_{18}OCH_3$, serum values of 20 nmol/ml are much smaller, which is probably the result of a much faster degradation of $ET_{18}OCH_3$ in different organs and tissues of the animals.

For the analysis of the organ distribution of HePC and $ET_{18}OCH_3$ after oral application, amounts of 75 μmol/kg were applied daily. The following results (fig. 7) were obtained with the HePTLC method described by Kötting et al. [this volume]. There was a dramatic difference between the biodistribution of HePC and $ET_{18}OCH_3$ as already indicated by the low serum levels of $ET_{18}OCH_3$. The tissue levels of $ET_{18}OCH_3$ were generally low. Values of < 50 nmol/g were determined. However, note the exception for adrenal glands with concentrations of 120 nmol/g.

In contrast, HePC shows much higher serum and organ levels. After the first week of treatment, HePC reached values of 900 nmol/g kidney (fig. 7a). Adrenal glands, spleen and small intestine contained HePC concentrations between 250 and 400 nmol/g. As demonstrated earlier with radiolabelled HePC, kidney, liver and lung are the favorite target organs for this compound regardless whether an i.v. or p.o. route of administration was chosen [20]. After 3 weeks, the HePC concentration is very high in kidney and reaches values of 1,390 nmol/g (fig. 7b). Traces of blood in the urine samples indicate that this concentration is toxic for this organ. HePC is also enriched significantly with values of 640 nmol/g in adrenal glands.

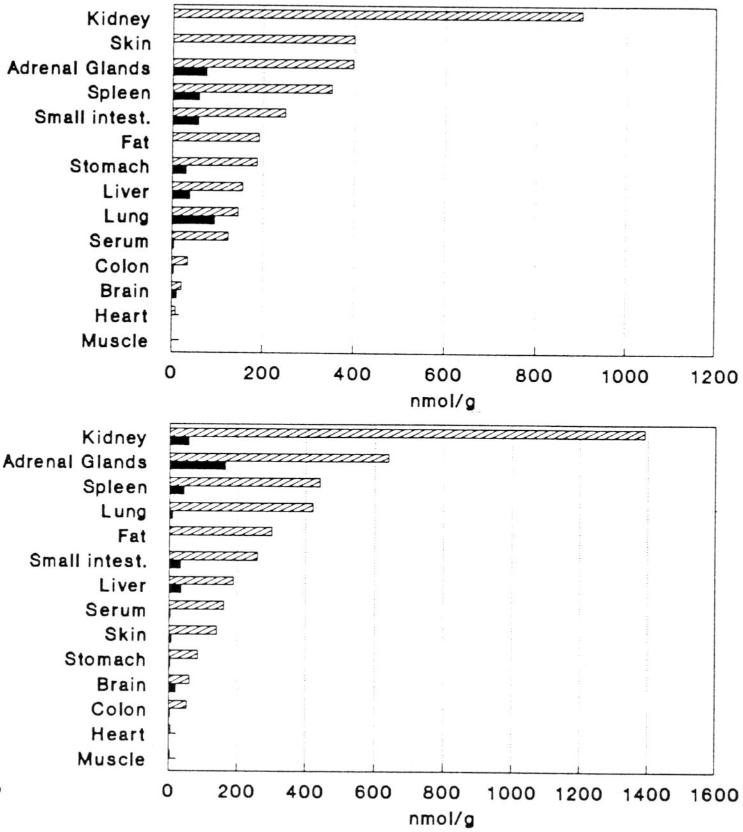

a *b*

Fig. 7. Biodistribution of HePC (□) and $ET_{18}OCH_3$ (■) in Wistar rats after p.o. application of the compounds for 1 week (*a*) and 3 weeks (*b*). Daily dosage was 75 μmol/kg, omitting Sundays. Animals were killed by CO_2 anesthesia and unbled by heart puncture. The tissue concentrations of HePC and $ET_{18}OCH_3$ were analyzed by thin-layer chromatography.

The ratio of the HePC concentration adrenal glands to kidney is 0.5 and constant over the time period studied. A rather large concentration of HePC is also detected in spleen. Values of 440 nmol/g are observed after 3 weeks.

In addition, skin and subcutanous fat were found to be another important depot of HePC. Since skin represents 20% of the body mass of the test animals, a value of 200 nmol/g is about 10% of the orally applied drug.

Concentration of HePC in the liver was similar to that of the serum with values of 140 nmol/g after 6 days and 170 nmol after the third week.

A remarkable enrichment in the lung was detected after the second week. The mean value changed from 145 to 420 nmol/g. Enrichment of the compound in body fat tissue (191 nmol/g) increased within 3 weeks to 300 nmol/g. The other tissues analyzed contained minor amounts of HePC, e.g. values of less than 10 nmol/g were found in colon, heart and muscle. In brain, the concentrations of HePC increased from 21 (1st week) to 64 nmol/g (3rd week).

In comparison, the $ET_{18}OCH_3$ concentrations in rat tissues are low (<100 nmol/g). After 1 week, highest amounts of $ET_{18}OCH_3$ were detected in lung (95 nmol/g), adrenal gland (75 nmol/g), spleen (59 nmol/g) and in small intestine (59 nmol/g). In liver, only 40 nmol/g were found. After 3 weeks of treatment, only adrenal gland (150 nmol/g) and kidney (55 nmol/g) showed an increase of $ET_{18}OCH_3$. In general, one can conclude from figure 7b, that there is no significant accumulation of $ET_{18}OCH_3$ in this 3-week study. This is consistent with data from an earlier investigation [20] with the radiolabeled compound. With respect to the quantitative absorption of HePC and $ET_{18}OCH_3$ (table 3), the low recoveries of the compounds found in rat tissues are probably explained by metabolic degradation of HePC [19] and $ET_{18}OCH_3$ [20].

The data presented here give evidence that there are striking differences in the pharmacological properties of the alkylphosphocholine HePC and the alkyllysolipid $ET_{18}OCH_3$. In spite of the quantitative absorption of both compounds in the small intestine of the rat, their serum concentrations differ as well as their binding to serum albumin and lipoproteins. Furthermore, the biodistribution of HePC and $ET_{18}OCH_3$ are qualitatively and quantitatively different. Generally, HePC concentrations in tissues are high in comparison to $ET_{18}OCH_3$ for all tissues. The better bioavailability of HePC can explain its increased therapeutic efficacy in in vivo experiments [8] and clinical trials [1], while their cytotoxicity in cell culture experiments [7] is almost identical.

References

1 Unger C, Eibl H, Nagel GA, Heyden HW von, Breiser A, Engel J, Stekar J, Peukert P, Hilgart P, Berger M: In Queisser W, Fiebig HH (eds): Hexadecylphosphocholine in the Topical Treatment of Skin Metastases, a Phase I Trial. Basel, Karger, 1989.

2 Unger C, Heyden HW von, Breiser A, Nagel GA: Etherlipids in the topical treatment of skin metastases. J Cancer Res Clin Oncol 1988;114:40.

3 Eibl H, Unger C: Hexadecylphosphocholine: A new and selective antitumour drug. Cancer Treat Rev 1990;17:233–242.

4 Kötting J, Fleer EAM, Unger C, Eibl H: Synthetische Alkyllysophospholipide als Antitumormittel-Strukturverwandte des 'Platelet Activating Factors'. Fat Sci Technol 1988;90:345–351.

5 Kötting J, Unger C, Eibl H: Substrate specificity of O-alkylglycerol monooxygenase (E.C. 1.14.16.5), solubilized from rat liver microsomes. Lipids 1987;22:831–835.
6 Neumüller W, Fleer EAM, Unger C, Eibl H: Enzymatic acylation of ether and ester lysophospholipids in rat liver microsomes. Lipids 1987;22:808–812.
7 Fleer EAM, Kim DJ, Nagel GA, Eibl H, Unger C: Cytotoxic activity of lysophosphatidylcholine analogues on human lymphoma Raji cells. Onkologie 1990;13:295–300.
8 Berger MR, Muschiol C, Schmähl D, Unger C, Eibl H: In Nagel GA, Sauer R, Schreiber HW (eds): Chemotherapeutische Studien zur Struktur-Wirkungs-Beziehung zytotoxischer Alkyllysophospholipide an chemisch induzierten Mammakarzinomen der Ratte. München, Zuckschwerdt, 1987.
9 Eibl H, Woolley P: A general synthetic method for enatiomerically pure ester and ether lyso-phospholipids. Chem Phys Lipids 1988;47:47–53.
10 Woolley P, Eibl H: Synthesis of enantiomerically pure phospholipids including phosphatidyl serine and phosphatidyl glycerol. Chem Phys Lipids 1988;47:55–62.
11 Kötting J, Marschner NW, Unger C, Eibl H: Membrane pertubation and hemolytic action of alkylphosphocholines and (alkyl-) lysolecithines (submitted).
12 Reman F, Demel R, De Gier J, Van Deenen LLM, Eibl H, Westphal O: Studies on the lysis of red cells and bimolecular lipid leaflets by synthetic lysolecithins, lecithin structural analogues. Chem Phys Lipids 1969;3:221–233.
13 Armstrong VW, Walli AK, Seidel D: Isolation, characterization, and uptake in human fibroblasts of an apo(a)-free lipoprotein obtained on reduction of lipoprotein(a). J Lipid Res 1985;26:1314–1323.
14 Parthasarathy S, Subbaiah PV, Ganguly I: The mechanism of intestinal absorption of phosphatidylcholine in rats. Biochem J 1974;140:503–508.
15 Charbonnier M, Grataroli R, Portugal H, Lafont H, Nalbone G: Subcellular distribution of lysophospholipase of rat intestinal mucosa. Biochem Cell Biol 1988;66:813–820.
16 Brindley DN: Metabolism of Triacylglycerols. Menlo Park, The Benjamin/Cummings Publishing Company, 1985.
17 Laritigue-Mattei C, Godeneche D, Chabard JL, Petit J, Berger JA: Pharmacokinetic study of ^{3}H-labelled lyso-PAF-acether II. Comparison with ^{3}H-labelled lyso-PAF-acether after intravenous administration in the rabbit and protein binding. Agents Actions 1984;15:643–648.
18 Spector AA: Fatty acid binding to plasma albumin. J Lipid Res 1975;16:165–179.
19 Breiser A, Kim, DJ, Fleer EAM, Damenz W, Drube A, Berger MR, Nagel GA, Eibl H, Unger C: Distribution and metabolism of hexadecylphosphocholine in mice. Lipids 1987;22:925–926.
20 Arnold B, Reuther R, Weltzien HU: Distribution and metabolism of synthetic alkyl analogs of lysophosphatidylcholine in mice. Biochem Biophys Acta 1978;530:47–55.
21 Neumüller W: Biosynthese von Phospholipiden: Über die Bedeutung aclierender Systeme in Rattenleber. Braunschweig, Technische Universität, 1988.

Jochem Kötting, Max-Planck-Institut für biophysikalische Chemie, Am Fassberg, D-W-3400 Göttingen (FRG)

Eibl H, Hilgard P, Unger C (eds): Alkylphosphocholines: New Drugs in Cancer Therapy. Prog Exp Tumor Res. Basel, Karger, 1992, vol 34, pp 143–152

Comparison of the Tissue Distribution of Hexadecylphosphocholine and Erucylphosphocholine

Ronald C. Reitz[a], *Jochem Kötting*[a], *Clemens Unger*[b], *Hansjörg Eibl*[a]

[a]Max-Planck Institute for Biophysical Chemistry, and
[b]Division of Hematology/Oncology, Department of Internal Medicine, University Hospital Göttingen, FRG

There are many antineoplastic agents, but the alkylphosphocholines represent a relatively new and exciting class of compounds, discovered by Eibl and Unger [1], which have been shown to possess both in vivo [2–3] and in vitro [3, 4] activity. These compounds represent the second generation of antitumor compounds and were originally designed to improve upon the alkyllysophosphatidylcholines in terms of physical properties and in terms of metabolism. The chance of metabolism of the alkylphosphocholines would be greatly reduced due to their more simple chemical structure. The lysophosphatidylcholines and their alkyl analogs have been known for quite some time to possess the ability to alter humoral and cell-mediated immune reactions [5, 6] as well as the growth of tumors in mice [5]. Further, the alkyl lysophosphatidylcholines have considerably higher metabolic stability than their ester analogs, and this permits better absorption of the undegraded molecule. However, there is measurable metabolism of these compounds [7], but at a sufficiently slow rate that accumulation of them is still allowed in the tissues. In addition, certain tissues seem to have a greater ability to accumulate these compounds compared to others [7], with kidney, liver and intestine being able to accumulate the greatest amounts. Recently, Breiser et al. [8] have shown that the alkylphosphocholine, hexadecylphosphocholine (HePC), also distributes selectively into kidney, liver and lung. They also suggested that a small amount of metabolism occurred by showing that radiotracer label in the choline moiety can be found in free choline from liver after an intravenous injection of radiolabelled HePC. Because this previous study [8] used tracer amounts of HePC via an i.v. injection, it was decided to reexamine the tissue distribution of HePC following a large oral bolus (40 μmoles). Further, it was decided to compare the distribution of HePC with that of erucylphosphocholine (EPC), an alkylphosphocholine with considerably different physical properties.

Materials and Methods

Hexadecylphosphocholine (HePC) and erucylphosphocholine (ErPC) were prepared by the procedure described by Eibl and Woolley (9) and Eibl and Westphal [10] for the conversion of fatty alcohols into the respective alkylphosphocholines. 1-Oleyl-2-hexadecyl-rac-glycero-3-phosphocholine (dialkyl-GPC) was prepared as described earlier [9, 11]. Two groups of 12 female Wistar rats (205–225g) per group were intubated intragastrically with 40 μmoles per rat of the alkylphosphocholines. One group was for the HePC and the other for ErPC. The HePC was in a total volume of 0.4 ml, and the ErPC was in a total volume of 1.0 ml, and the animals were housed in groups of three after the administration of the alkylphosphocholines. The animals were killed by CO_2 anesthesia after 2, 4, 6 and 10 days. A third group of animals was intubated with either ErPC (3 animals) or dialkyl-GPC (3 animals), 40 μmoles, and the total feces from each animal collected at 24 and 48 hrs. In addition, the feces remaining in the large intestine after 48 hours were also collected. The animals were intubated and/or killed between 9:00 and 10:30 am, and various tissues were removed, weighed and immediately frozen at −70 °C until extraction by the procedure of Bligh and Dyer [12]. The feces were also extracted by this same procedure, but in order to facilitate the chromatography and detection, these lipids were subjected to mild alkaline hydrolysis (0.25 N KOH in methanol heated at 50–60 °C for 5 minutes) and reextracted before TLC. Both ErPC and dialkyl-GPC are stable against alkaline hydrolysis. Thus, there were no acyl compounds to interfere with the densitometry. The lipids were dissolved in chloroform : methanol : water (30 : 60 : 8 per vol.) or chloroform : methanol (2 : 1 v/v) and spotted on HPTLC plates without fluorescent indicator (Kieselgel 60, No. 5641, E. Merck, Darmstadt, FRG). They were developed in chloroform : methanol : triethylamine : water (30 : 35 : 34 : 8 per vol.) for analysis of HePC and dialkyl-GPC and (30 : 50 : 34 : 8 per vol.) for ErPC. They were then dried, charred after dipping in 10% $CuSO_4$ in 8% H_3PO4, and quantitated by densitometry following the procedure of Kotting et al. [this volume]. Pure HePC, dialkyl-GPC and ErPC were used as quantitative standards on each plate depending upon which compound was to be analyzed.

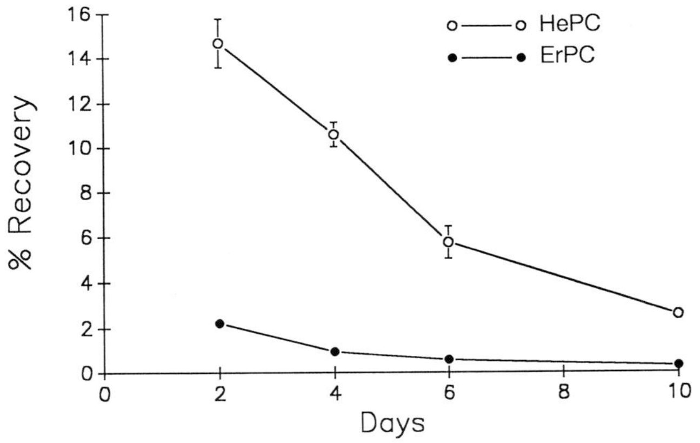

Fig. 1. Recovery of HePC and ErPC. The total tissue recovery from 10 tissues was used to calculate the percentage recovery of either HePC or ErPC. 40 μmoles of each compound was given intragastrically and the animals were killed at 2, 4, 6 and 10 days after this single dose and the content of each compound was determined. Each value represents the mean ± SD from 3 animals.

Results

Ten tissues were extracted for total lipids and analyzed for their content of either HePC or ErPC. The tissues were liver, lung, kidney, spleen, adrenal gland, serum, stomach, duodenum, jejunum and ileum. Figure 1 shows that the percent recovery of HePC was considerably greater than ErPC at all 4 days of the experiment. At day 2 after the intubation, the total recovery of HePC was 14.7% and ErPC was only 2.2%. The ten tissues extracted accounted for 20.0–20.8% of the total mass of the animals. If there were even distribution into all tissues, these data suggest that there is a loss of about 22% of the HePC and 90% of the ErPC between day 0 and day 2. To further check the loss of ErPC, an additional set of 3 animals was intubated with 40 μmoles of ErPC and the feces collected at 24 and 48 hrs. The feces remaining in the large intestine after 48 hrs were also collected. These three samples of feces were extracted and analyzed. As a control, 40 μmoles of 1-oleyl-2-hexadecyl-glycero-3-phosphocholine, Dialkyl-GPC, were intubated into an additional set of 3 animals, the feces collected and the samples assayed in a similar manner to ErPC. Table 1 shows that after 1 day 3% of the total amount of ErPC could be extracted from feces. During the next 24 hrs, an additional 4% was extracted, and 0.8% was found in the feces remaining in the large intestine of the animals. Thus, after 2 days, about 8% of the total dose of ErPC can be found in the feces. On the other

Table 1. Recovery of ErPC and Dialkyl-GPC in Feces

	ErPC Recovered		Dialkyl-GPC	
	nmoles	% Recovery	µmoles	% Recovery
Day 1	1201 ± 163	3.0	41.0 ± 4.8	102.5
Day 2	1593 ± 252	4.0	0.83 ± 0.51	2.1
Day 2 (Internal*)	323 ± 82	0.8	0.0	0.0

Feces were collected at 24 hr intervals, and after 48 hrs, the animals were killed and the feces present in the large intestine were collected, extracted and analyzed for ErPC or Dialkyl-GPC. Each value represents the mean ± SD from 3 animals.
(*Feces from Large Intestine)

hand, the dialkyl-GPC was totally excreted from the animals during the first 24 hrs.

To simplify the visualization of the data, the distribution of HePC and ErPC into the different tissues has been divided into two groups. The first group represents the tissues which should be the first to come in contact with the two alkylphosphocholines. These are the stomach, duodenum, jejunum, ileum and serum. The distribution and the rate of loss of these two compounds from these 5 tissues is shown in Fig. 2 and 3. In both cases, the duodenum contained the most HePC or ErPC, and the stomach contained the least. The relative distribution in the other tissues was also quite similar for each of the two compounds. The most remarkable difference was the magnitude of the concentration of HePC compared to ErPC in the various tissues. There was between 2 and 6 fold more HePC per gram tissue than ErPC. After 10 days, the duodenum has lost 87% of its ErPC but only 77% of the HePC. The stomach has lost 92% of its ErPC and only 84% of its HePC. The ileum is similar to the stomach having lost 92% of its ErPC and only 84% of its HePC. In the jejunum, both compounds were lost to a similar extent, about 90%. Thus, in this group of tissues, ErPC seems to be lost at a slightly faster rate.

Fig. 4 and 5 show the distribution and loss of HePC and ErPC from other internal organs. It is interesting to compare lung and kidney. With respect to HePC, kidney accumulated the most of any tissue and this was 3 times more than the lung tissue which contained the second most concentrated amount of HePC. On the other hand, ErPC accumulated about 3 times more in lung than in the kidney. It is also interesting to note that both liver and adrenal accumulated a higher concentration of ErPC than kidney. Also these tissues accumulated concentrations of HePC from 3 to 62 fold higher than ErPC with the kidney being the most selective. Both

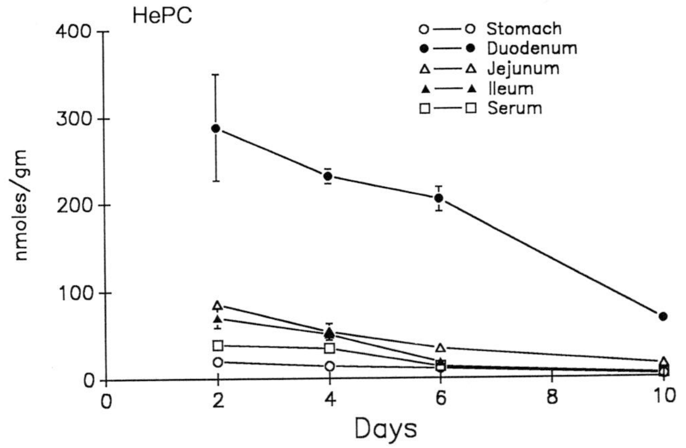

Fig. 2. Concentration of HePC in Gastrointestinal Tissues. After 40 μmoles of HePC were given orally, the concentration was measured in 5 GI tissues and serum at 2, 4, 6 and 10 days. Each value represents the mean ± SD from 3 animals.

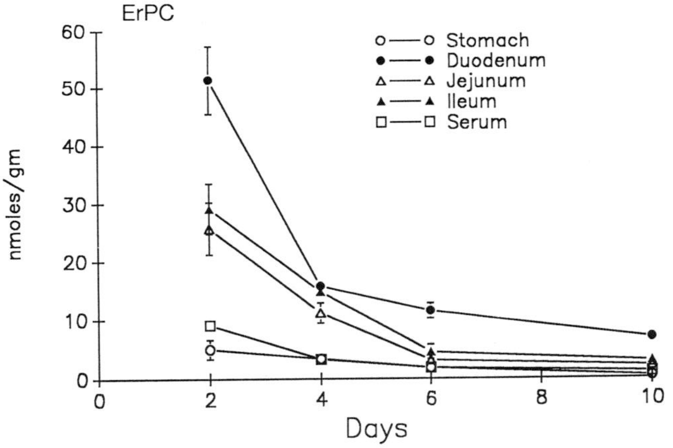

Fig. 3. Concentration of ErPC in Gastrointestinal Tissues. After 40 μmoles of ErPC were given orally, the concentration was measured in 5 GI tissues and serum at 2, 4, 6 and 10 days. Each value represents the mean ± SD from 3 animals.

kidney and spleen retained the most HePC after 10 days, with the loss being only about 70%. Lung seemed to have quite different kinetics for the loss of the ErPC molecule. About 86% of the ErPC was lost by day 4 compared to day 2, and this further declined to 96% by day 10. Only 43% of the HePC was lost from lung at day 4 compared to day 2. Another

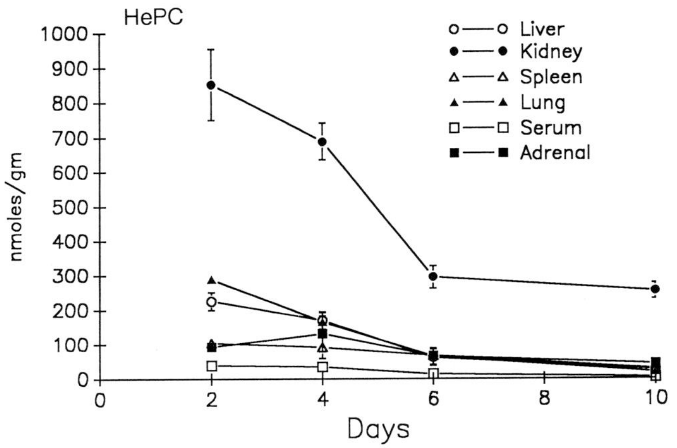

Fig. 4. Concentration of HePC in Internal Tissues. After 40 μmoles of HePC were given orally, the concentration was measured in 5 Internal tissues and serum at 2, 4, 6 and 10 days. Each value represents the mean ± SD from 3 animals.

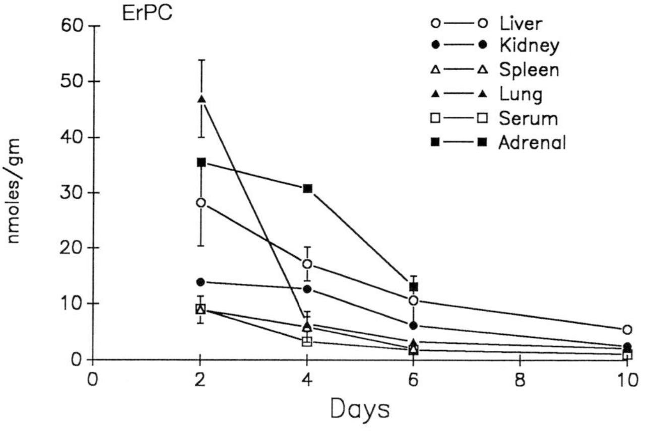

Fig. 5. Concentration of ErPC in Internal Tissues. After 40 μmoles of ErPC were given orally, the concentration was measured in 5 Internal tissues and serum at 2, 4, 6 and 10 days. Each value represents the mean ± SD from 3 animals.

interesting point was that the adrenal accumulated the second highest concentration of ErPC of any of these internal tissues. Overall only lung and duodenum accumulated higher concentrations of ErPC than adrenal.

Table 2 presents the percent distribution in the 10 tissues based upon per gram tissue rather than based on the whole tissue. The distribution of

Table 2. Tissue Distribution of Alkylphosphocholines

	HePC	ErPC
Primary	Kidney (41%)	Duodenum (20%) Lung (19%)
Secondary	Lung (14%) Duodenum (14%) Liver (11%)	Adrenal (14%) Liver (11%) Ileum (11%) Jejunum (10%)
Tertiary	Spleen (5%) Adrenal (5%) Jejunum (4%) Ileum (3%) Serum (2%)	Kidney (5%) Spleen (4%) Serum (3%)

Distribution figured 2 days after intubation. Percentages are based upon the total amount of compound recovered on a per g weight basis.

each compound can be grouped into three categories: primary, secondary and tertiary. HePC showed a very distinct preference for accumulating in the kidney, while ErPC accumulated mainly in duodenum and lung. The HePC in the kidney was about 3 fold more concentrated than in the secondary tissues of lung, duodenum and liver. The ErPC was only 2 fold more concentrated in duodenum and lung compared to the secondary tissues of adrenal, liver, jejunum and ileum. Comparing the primary and tertiary tissues, HePC was about 10 fold more concentrated in its primary tissue, and ErPC was only 4–5 fold more concentrated.

Discussion

Our data strongly suggest that the absorption of HePC is considerably greater than that for ErPC. Two days after the intubations, we can account for about 15% of the total dose of HePC from 10 tissues which amount to 20% of the total body weight of the animal. We recovered only 1% of the total dose of HePC in the feces and less than 1% in the urine after 24 hrs from animals intubated with only 5 μmoles. These data indicate almost complete absorption of HePC. Breiser et al. (8) presented evidence that the liver is capable of metabolizing HePC by showing that 37% of an i.v. administered dose of 476 nmoles was found either as diacylphosphatidylcholine (2%), phosphocholine (3%) or choline (32%) after 24 hrs. Thus, our failure to show complete recovery after 48 hrs, can probably be explained by metabolism in the liver as suggested by Breiser et al. [8]. However, the loss of 90% of the administered dose of ErPC strongly suggests another mechanism for loss. There is no available data regarding the metabolism of ErPC, but we recovered almost 8% of the total dose in feces. The fact that 8% ErPC was recovered after 2 days in the feces indicates that this longer chain alkylphosphocholine is poorly absorbed compared to HePC. In combination, we can account for 10% of the total ErPC dose. When one considers that the tissues we extracted amount to 20% of the total weight of the animal, then the 10% recovery represents about 50% of what we might expect. The CMC for HePC is 8.0×10^6 [see Kötting et al., this volume], and ErPC forms lamellar bodies not micelles. Thus, it is possible that the physical form contributes significantly to the absorption characteristics of these two compounds.

The transfer of phosphatidylcholine from blood to the cells of other tissues seems to require that PC be converted into lysophosphatidylcholine first [13], and it is generally accepted that phosphatidylcholine is only absorbed in the GI tract as the lysophospholipid [14–16]. Using double labeled phosphatidylcholine, Scow et al. [14] showed that phosphatidylcholine was hydrolyzed to lysophosphatidylcholine before absorption from the GI tract. The complete recovery of the dialkyl phosphatidylcholine (Table 1) strongly supports this concept. However, because only 8% of the total amount of ErPC was excreted in two days and because we could not find HePC in feces, the absorption characteristics of the alkylphosphocholines are quite different compared to the dialkyl-GPC. The longer chain length of ErPC compared to HePC seems to give the molecule more of the properties of the dialkyl-GPC, while the increased water solubility of HePC seems to give this molecule more of the properties of a detergent like lysophosphatidylcholine. ErPC has more lamellar character and less detergent character compared to HePC, yet there was still some absorption of ErPC. The most difficult task is to explain the 25% loss of HePC and the 75% loss in ErPC. The selective distribution of each of these compounds in the various tissues suggests that there may be a tissue that we have not measured which has an inordinate amount of either HePC or ErPC.

The rate of elimination from each tissue must also play a role in the concentrations observed at the different time points. Overall, ErPC can be calculated to be lost at about 0.096 nmols/day and HePC at 0.60 nmoles/day over the 10 day experimental period. This amounts to a 6.25 fold higher loss of HePC on a daily basis. During the 10 day experimental period, both ErPC and HePC decreased by about the same percentage (88% and 82%, respectively). Because Breiser et al. [8] have shown that the liver can metabolize HePC, some of this loss may be related to metabolism especially in liver.

Probably one of the most interesting observations is the unique tissue distribution of the two compounds. Arnold et al. [7] have reported that other alkyl phospholipids have selective tissue distribution. They observed that 1-octadecyl-2-methyl-rac-glycero-3-phosphocholine and 1-hexadecyl-propanediol-3-phosphocholine have similar distributions preferring mainly the intestine (16%, 18%) and liver (14%, 14%) with secondary preferences to kidney (7%, 6%) and blood (6%, 2%); however, 1-dodecyl-propane-diol-3-phosphocholine had a preference for kidney (20%) with a secondary preference for liver (11%). The HePC had a very distinct preference for internal tissues with kidneys having the greatest preference for this compound. This is quite similar to the distribution of HePC in these same tissues after an i.v. injection of radiotracer HePC [8]. By increasing the chain length and unsaturation, the ErPC compound, on the other hand, seemed to prefer the tissues of the GI tract plus the lung and the liver. Thus, the hydrophobic end of the molecule seems to play a role in tissue specificity. This is also suggested by the data of Arnold et al. [7] for the alkyl analogs of lysphosphatidylcholines. It is quite clear that the difference in physical properties of these two compounds, HePC and ErPC, result in very different absorption characteristics from the GI tract and very different tissue distributions after absorption.

References

1 Eibl H, Unger C: Hexadecylphosphocholine, a new and selective antitumor drug. Cancer Treat Rev 1990;17:233–242.

2 Muschiol, C., Berger, M.R., Schuler, B., Scherf, H.R., Garzon, R.T., Zeller, W.J., Unger, C., Eibl, H.J. and Schmahl, D. (1987) "Alkyl Phosphocholines: Toxicity and Anticancer Properties." Lipids 22, 930–934.

3 Unger, C., Damenz, W., Fleer, E.A.M., Kim, D.J., Breiser, A., Hilgard, P., Engel, J., Nagel, G., and Eibl, H. (1989) "Hexadecylphosphocholine, A New Ether Lipid Analogue", Acta Oncologica 28, 213–217.

4 Fleer, E.A.M., Kin, D.J., Nagel, G.A., Eibl, H., Unger, C. (1990) "Cytotoxic Activity of Lysophosphatidylcholine Analogues on Human Lymphoma Raji Cells." Onkologie 13, 295–300.

5 Munder, P.G., Weltzien, H.U. and Modolell, M. (1977) "Lysolecithin Analogues: A New Class of Immunopotentiators." in Immunopathology, VIIth Int. Symposium, Bad Schachen 1976, (Miescher, P.A., ed.) pp. 411–424, Schwabe and Co. Publishers, Basel.

6 Stannegard, O. and Roupe, G. (1976) "Adjuvant Effect of Lysolecithin analogs on the Development of Contact Sensitivity in Mice." Int. Arch. Allergy 51, 198–205.

7 Arnold, B., Reuther, R. and Weltzien, H.U. (1987) "Distribution and Metabolism of Synthetic Lysophosphatidylcholine Analogs in Mice." Biochim. Biophys. Acta 530, 47–55.

8 Breiser, A., Kim, D.J., Fleer, E.A.M., Damenz, W., Drube, A., Berger, M., Nagel, G.A., Eibl, H. and Unger, C. (1987). "Distribution and Metabolism of Hexadecylphosphocholine in Mice." Lipids 22, 925–926.

9 Eibl, H. and Woolley, P. (1988) "A General Synthetic Method for Enantiomerically Pure Ester and Ether Lysophospholipids." Chem. Phys. Lipids 47, 63–68.
10 Eibl, H. and Westphal, O. (1967) "Synthesen von Cholinphosphatiden, II." Liebigs Ann. Chem. 709, 231–233.
11 Eibl, H. and Woolley, P. (1986) "Synthesis of Enantiomerically Pure Glyceryl Esters and Ethers. I. Methods Employing the Precursor 1,2-Isopropylidene-sn-Glycerol." Chem. Phys. Lipids 41, 53–63.
12 Bligh, E.G. and Dyer, W.J. (1959) "A Rapid Method of Total Lipid Extraction and Purification." Can. J. Biochem. Physiol. 37, 911–917.
13 Stein, Y. and Stein, O. (1966) "Metabolism of Lysolecithin, Phosphatidylethanolamine and Lecithin in the Rat." Biochim. Biophys. Acta 116, 95–197.
14 Scow, R.O., Stein, Y. and Stein, O. (1967) "Incorporation of Dietary Lecithin and Lysolecithin into Lymph Chylomicrons in the Rat." J. Biol. Chem. 242, 4919–4924.
15 Parthasarathy, S., Subbaish, P.V. and Ganguly, J. (1974) "The Mechanism of Intestinal Absorption of Phosphatidylcholine in Rats." Biochem. J. 140, 503–508.
16 Nilsson, A. (1968) "Intestinal Absorption of Lecithin and Lysolecithin by Lymph Fistula Rats." Biochim. Biophys. Acta 152, 379–390.

Ronald C. Reitz, Department of Biochemistry, University of Nevada,
Reno, NV 89557 (USA)

Eibl H, Hilgard P, Unger C (eds): Alkylphosphocholines: New Drugs in Cancer Therapy. Prog Exp Tumor Res. Basel, Karger, 1992, vol 34, pp 153–159

Hexadecylphosphocholine in the Topical Treatment of Skin Metastases in Breast Cancer Patients

C. Unger[a], *H. Sindermann*[b], *M. Peukert*[b], *P. Hilgard*[b], *J. Engel*[b], *H. Eibl*[c]

[a]Division of Hematology/Oncology, Department of Internal Medicine, University Hospital Göttingen; [b]ASTA Pharma AG, Frankfurt/Main; [c]Max-Planck Institute for Biophysical Chemistry, Göttingen, FRG

Introduction

Treatment of skin metastases in breast cancer patients has often presented a severe problem in medical oncology, predominantly in patients which are refractory to conventional treatment such as surgery, radiation or systemic therapy. Moreover, in a multitude of patients skin involvement or local recurrence may represent the only site of tumor manifestation for a long time. The patients psychological situation may be additionally complicated due to visibility of the tumors. For this reason topical application of a cytotoxic drug could be a way to treat metastases of the skin and to control tumor growth by reducing or preventing major systemic adverse effects.

General Requirements for Percutaneous Absorption

Topical treatment of skin metastases in breast cancer patients requires drugs which are active against this tumor type. Furthermore, such compounds should cause only limited local adverse effects such as inflammation or injury at the site of application. A number of factors determine whether a cytotoxic drug will penetrate the skin. Among these, the chemical structure is of importance, particularly with respect to ionization. Lipid solubility which is expressed as oil/water partition coefficient is also an important factor. Due to the heterogeneous composition of the stratum corneum it has been shown in many studies that molecules with a partition coefficient in the region of unity penetrate most widely [1]. That is, compounds which are good penetrants have a well-balanced water and lipid solubility.

Hexadecylphosphocholine as a Suitable Candidate

Alkylphosphocholines represent a new group of antitumor compounds which were developed in our laboratory [2]. Among these, hexadecylphosphocholine (HePC), a representative of the alkylphosphocholines, seemed to be a good candidate for topical treatment. Due to its amphiphilic behavior HePC was expected to possess favorable penetration characteristics. Moreover, a pronounced in vivo antitumor activity has been demonstrated in chemically induced, autochthonous rat mammary carcinomas [3, 4]. In particular, topically applied HePC has been shown to induce complete tumor regressions of subcutaneously inocculated tumor cells in nude mice [see Hilgard et al., this volume]. Thus, HePC was the first alkylphosphocholine selected for a topical treatment of skin metastases in breast cancer patients in clinical trials.

Short-Chain Alkylglycerols and Drug Penetration

To increase the penetration in the skin, HePC was dissolved into a mixture of short-chain alkylglycerols and water. Short-chain alkylglycerols are amphiphilic molecules. Dependent on the chain length of the alkyl residue they are differently soluble in aqueous systems. Whereas alkylglycerols with a chain length of C_1 to C_5 are water soluble, the solubility of compounds with longer alkyl chains is restricted. Short-chain alkylglycerols possess favorable solvent properties for lipophilic molecules and facilitate drug penetration through biological membranes. This was demonstrated for the first time on the blood-brain barrier in rats [5]. For instance, in the case of small molecules with a molecular weight up to 500 daltons, e.g. 5-fluorouracil, methotrexate or cyclophosphamide, the permeability of cytostatics was strongly increased if short-chain alkylglycerols, up to a final concentration of about 0.3 mol/l were added to the injected solution. Under these conditions the 'Brain-Uptake Index' (BUI) reached values up to about 50% (cyclophosphamide) (fig. 1), depending on lipid solubility and molecular dimension of the respective test compounds and the alkyl chain length of the glycerol derivatives (table 1).

A second favorable characteristic of short-chain alkylglycerols is their bactericidal and fungicidal activity. Here, too, the longer chain alkylglycerols exhibit a greater antimicrobial effect (fig. 2).

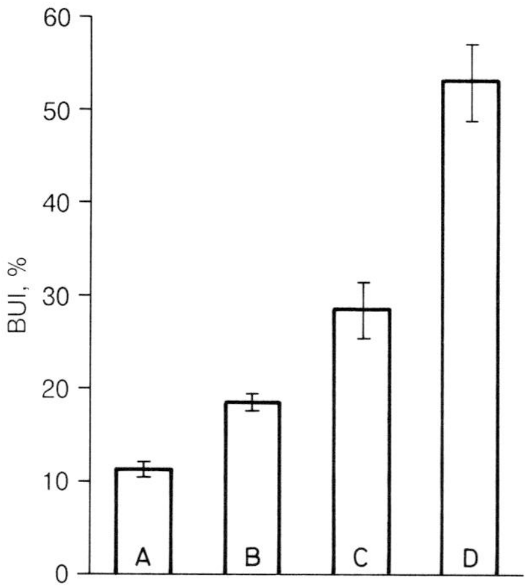

Fig. 1. BUI values of ^{14}C-cyclophosphamide: basal BUI (A) and BUI values in the presence of iso-osmolar concentrations (0.3 *M*) of 1-O-propylglycerol (B), 1-O-butylglycerol (C) and 1-O-pentylglycerol (D). n = 3–6 experiments ±SD.

Table 1. BUI values and oil/water partition coefficients (p) of cytostatics and glycerol

Test substance	MW	log p	BUI values, %	
			without PG	with PG
Sucrose	342	–3.00	3.3 ± 1.3	–
5-Fluorouracil	132	–2.89	3.7 ± 1.0	13.0 ± 0.1
Methotrexate	455	–2.44	5.3 ± 0.7	26.2 ± 5.8
Glycerol	92	–2.39	3.9 ± 1.9	27.4 ± 2.7
Cyclophosphamide	262	–1.03	11.4 ± 0.4	53.5 ± 6.1
Doxorubicin	544	–1.00	2.4 ± 0.4	–
Etoposide	589	–0.62	3.2 ± 0.8	17.8 ± 3.2
Danuorubicin	528	–0.30	1.4 ± 0.3	14.3 ± 1.2
Bleomycin	1,400	–1.89	2.3 ± 0.6	5.7 ± 1.8
Peplomycin	1,400	–0.74	4.2 ± 0.4	13.8 ± 1.5
Vincristin	923	–0.47	1.0 ± 0.2	–
Mitoxanthrone	625	+0.13	1.8 ± 0.1	25.1 ± 2.0
Vinblastine	909	+0.99	1.5 ± 0.2	41.4 ± 4.8

The penetration of radiolabeled cytostatics through the blood-brain barrier of rats was measured by the method of Oldendorf [Oldendorf WH: Brain Res 1970;24:372–376]. The BUI value for sucrose (3.3%) corresponds to the blank value of the method and signifies no passage across the barrier. 1-O-pentylglycerol (PG) was used in iso-osmolar concentration (0.3 *M*). n = 4–6 experiments ±SD.

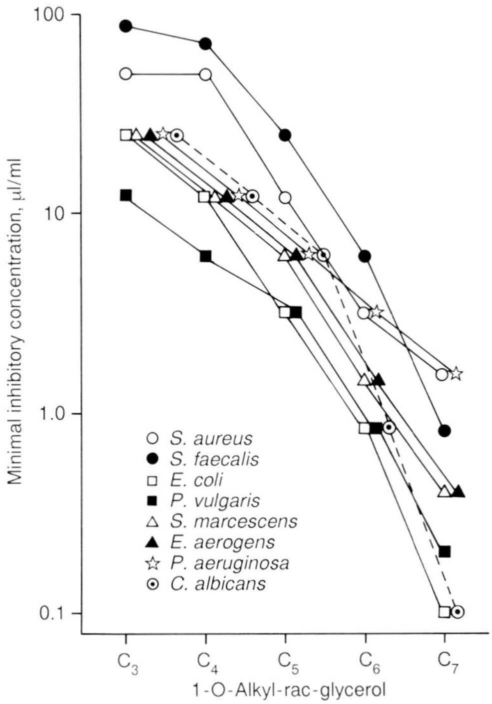

Fig. 2. Susceptibility of infecting microorganisms to short-chain alkylglycerols with a chain length of C_3 to C_7. Minimal inhibitory concentrations were determined by serial dilutions of each alkyglycerol to be tested in IST-Bouillon adding an inoculum of 10^5/ml of the infecting organisms. Incubations were carried out for 18 h at 37 °C. These experiments were performed by Dr. Ansorg (University Hospital, Göttingen).

Composition of the HePC/Alkylglycerol/Water Mixture

For topical application, HePC was dissolved in a mixture of 1-O-alkylglycerols consisting of water, 1-O-propylglycerol, 1-O-hexylglycerol and 1-O-nonylglycerol in a ratio of 2:2:1:1 (by weight) (table 2). The proportions of the different alkylglycerols and water were chosen to reach a homogeneous solution. The obtained mixture is slightly viscous and can easily be applied on the affected skin. The final concentration of HePC in the solution was 20, 40, 60 and 80 mg HePC/ml.

Table 2. Composition of the HePC/alkyglycerol/water mixture

Substance	% (weight)
1-O-Nonylglycerol	17
1-O-Hexylglycerol	17
1-O-Propylglycerol	33
Water	33

20, 40, 60 or 80 mg HePC were dissolved in 1 g of the alkylglycerol/water solution.

Table 3. Patient characteristics

Patients, n	
Total	25
Evaluable	24
Age range, years (median)	39–85 (55)
Pretreatment	
Surgery	22
Radio-, chemo- and hormone therapy	18
Radio- and chemotherapy	3
Radio- and hormone therapy	2
Radiotherapy only	1

Clinical Investigations

Patients and Methods

Twenty five patients with metastatic breast cancer and progressive skin involvement were entered into the clinical trial, 24 of whom were evaluable. The patients had been extremely pretreated, as indicated in table 3. No other antiproliferative treatment was administered during the topical HePC treatment.

The study protocol was reviewed and approved by the local ethical committee. After informed consent, HePC (20 mg/ml) was applied once a day for 5 consecutive days. In the absence of side effects, the therapy was continued for 2 weeks with application twice daily. In the case of tolerance, the dosage was escalated progressively every 2 weeks by increments of 20 mg HePC/ml up to 80 mg/ml. To ensure adequate evaluation of response and toxicity the therapy was maintained for at least 8 weeks. Topical treatment was discontinued in the case of tumor progression, unacceptable local intolerance or as soon as other treatment modalities had to be given priority. In the case of no change or partial remission the treatment was continued until exclusion criteria became effective.

Results

Evaluation of the individual dose level of HePC was based on the following numbers: 20 mg/ml in 24 patients, 40 mg/ml in 13 patients,

Table 4. Results with HePC in the topical treatment of skin metastases in metastatic breast cancer (n = 24)

Complete remission	4	(1–6 months)
Partial remission	3	(2–12 months)
No change	8	(2–31 months)
Progressive disease	9	

60 mg/ml in 11 patients and 80 mg/ml in 5 patients. The minimum length of treatment was 3 weeks, maximum 128 weeks (median 18 weeks). The treatment was well tolerated by all patients. Occasional itching was reported, but in no case had the therapy to be discontinued due to local intolerance. In 4 of 24 patients, a slight erythema (WHO grade 1) was noted; in 1 patient the erythema was WHO grade 2/3. After prolonged application, scaling and dryness of the treated area was observed with regularity. In 3 cases, skin biopsies were taken from these areas after 3–12 months of therapy. The histological results revealed normal skin architecture with a slight increase in collagen formation in the upper dermis. During dose escalation, even 80 mg/ml was well tolerated. However, application to a non-pretreated skin area resulted in a marked skin erythema which was reversible by lowering the HePC dose to 60 mg/ml. In particular, it should be stressed that neither hematotoxic nor other blood chemistry changes could be observed. As indicated in table 4, we noted 4 complete remissions (1–6 months), 3 partial remissions (2–12 months), 8 no changes (2–31 months) and 9 progressive diseases. The 4 patients with complete remission had few to multiple flat skin metastases up to 2 cm in diameter and responded already to 20 mg HePC/ml (2 cases) as well as to 40 and 60 mg HePC/ml (1 case on each dose level). The partial remissions were obtained with 20 and 60 mg HePC/ml, respectively. Ulcerating tumors responded to topical treatment with progressive clearing of the ulcerated area and a reduction of smelling. Starting at the edge of the ulcerations a re-epithelialisation was noted.

A phase II trial of HePC 60 mg/ml in patients with skin metastasized breast cancer is currently ungoing in the phase II study group of the German Society of Medical Oncology (AIO). In an interim evaluation the results appear to be similar to those obtained in the phase I trial.

References

1 Stüttgen G, Schäfer H: Die Permeabilität der Haut; in: Funktionelle Dermatologie, Grundlagen der Morphokinetik, Pathophysiologie, Pharmakoanalyse und Therapie von Dermatosen Berlin Springer, 1974, pp 353–376.

2 Eibl H, Unger C: Hexadecylphosphocholine, a new and selective antitumor drug. Cancer Treat Rev 1990;17:233–242.

3 Muschiol C, Berger MR, Schuler B, Scherf HR, Garzon FT, Zeller WJ, Unger C, Eibl H, Schmähl D: Alkylphosphocholines: Toxicity and anticancer properties. Lipids 1987;22:930–934.

4 Hilgard P, Stekar J, Voegeli R, Engel J, Schumacher W, Eibl H, Unger C, Berger MR: Characterization of the antitumor activity of hexadecylphosphocholine (D 18506). Eur J Cancer Clin Oncol 1988;24:1457–1467.

5 Unger C, Eibl H, von Heyden HW, Krisch B, Nagel GA: Blut-Hirnschranke und Penetration von Zytostatika. Klin Wochenschr 1985;63:565–571.

C. Unger, Division of Hematology/Oncology, Department of Internal Medicine, University Hospital Göttingen, D-W–3400 Göttingen (FRG)

Eibl H, Hilgard P, Unger C (eds): Alkylphosphocholines: New Drugs in Cancer Therapy. Prog Exp Tumor Res. Basel, Karger, 1992, vol 34, pp 160–169

Topical Application of Hexadecylphosphocholine in Patients with Cutaneous Lymphomas

Reinhard Dummer[a,c] *Jürgen Röger*[a], *Thomas Vogt*[a], *Jürgen Becker*[a], *Hiltrud Hefner*[a], *Herbert Sindermann*[b], *Günter Burg*[a,c]

[a]Department of Dermatology, University of Würzburg Medical School, Würzburg;
[b]Clinical Cancer Research, ASTA Pharma AG, Frankfurt, FRG;
[c]Department of Dermatology, University of Zürich Medical School, Zürich, Switzerland

Introduction

Cutaneous lymphomas comprise a heterogeneous group of clonal lymphoproliferative disorders originating in the skin. They are differentiated analogous to the structure of the lymphatic systems in cutaneous B and T cell lymphomas [1, 2]. Recently, the development of immunhistochemistry allowed a more sophisticated classification of these neoplasms (table 1) [3].

Another disease closely related to cutaneous T cell lymphomas is lymphomatoid papulosis. The cutaneous lesions of the latter lymphoproliferative disorder consist of CD-30-positive blastic T lymphocytes. The papulonodular eruptions resemble pityriasis lichenoides et varioliformis acuta (Mucha-Haberman disease) clinically, but lymphoma histologically [4]. A clonal rearrangement of the T cell receptor beta-chain had been demonstrated [5]. The clinical course is rather benign. The disease is recurrent, but sometimes self-healing. However, there is an as yet undefined relationship to Hodgkin's lymphoma [6].

Low-grade peripheral T cell lymphomas represent the majority of cutaneous lymphomas. The most common diseases in this group include mycosis fungoides and its leukemic counterpart, the Sézary syndrome. The first symptoms of these cutaneous lymphomas appear in the skin. In the early stages, flat lesions such as macules and patches or slightly elevated lesions such as papules or plaques are dominating, whereas in the more advanced stages also larger lesions such as nodules or tumors develop [1, 2].

The early stage skin lesions are often misdiagnosed as chronic contact dermatitis or eczema, because they present only nonspecific histological features such as spongiosis and sparse infiltrate of mononuclear cells.

Table 1. Histological classification of malignant cutaneous lymphomas

Cutaneous T cell lymphoma
A Prethymic and thymic T cell lymphoma
Lymphoblastic
B Peripheral T cell lymphoma
I Low grade
1 Chronic lymphatic or polymorphocytic leukemia
2 Small cerebriform (mycosis fungoides, Sézary syndrome)
3 Lymphoepitheloid (Lennert's) lymphoma
4 Angioimmunoblastic lymphoma
5 T-zone lymphoma
6 Pleomorphic, small cell lymphoma
II High-grade
1 Pleomorphic, medium and large cell lymphoma
2 Immunoblastic lymphoma
3 Large cell anaplastic lymphoma
III Unclassifiable
Cutaneous B cell lymphoma
I Low grade
1 Lymphocytic lymphoma
2 Lymphoplasmacytoid lymphoma
3 Centrocytic lymphoma
4 Centroblastic-centrocytic lymphoma
II High-grade
1 Centroblastic lymphoma
2 Immunoblastic lymphoma
3 Lymphoblastic lymphoma
III Unclassifiable
Lymphomatoid papulosis

However, repeated biopsies may help to identify characteristic histological criteria for the diagnosis of cutaneous T cell lymphoma. They include an epidermotropic, band-like mononuclear infiltrate, or epidermal collections of lymphatic cells (Pautrier's microabscess) [1, 2].

Cutaneous lymphomas are usually slowly progressing. There are often years between the first appearance of skin lesions and the definite diagnosis. Lymph nodes and other organs are involved in advanced stages, after the T cells have lost their epidermotropic distribution pattern and start to form nodules or tumors.

There have been attempts to apply aggressive treatment modalities such as high-dose chemotherapy or radiation. The long-term results, however, were disappointing. A recently published randomized clinical trial failed to show an increased survival for a high-dose chemotherapy plus radiotherapy compared to stage-adapted therapy [7]. As a consequence, the treatment of cutaneous lymphomas is palliative. A topical therapy is preferred as long as possible. For the early stages, local application of steroids such as betametasone or photochemotherapy (psoralen plus UV-A) are used [8]. Advanced stages require the topical application of cytotoxic drugs such as carmustine (BCNU) or mechlorethamine (NH2) [9]. Interferon-alpha seems to be an effective cytokine for the therapy of cutaneous T cell lymphoma [10]. Systemic involvement is treated by chemotherapy or radiation in case of single lesions.

However, long-term treatment with these drugs results in various dose-limiting side effects. Steroids induce atrophy of the dermis and epidermis [11]. Photochemotherapy is associated with an increased incidence of epidermal neoplasia such as basal cell carcinoma [12]. Finally, cutaneous T cell lymphomas often lose their responsiveness to these topical drugs. Therefore, therapeutic alternatives are required for cutaneous lymphomas. Epidermotropism and long-lasting restriction to the skin are pathophysiological properties which suggest further development of topical therapy. Since lymphatic tumor cells are suitable targets for phospholipid derivates [13, 14], this new group of cytotoxic drugs is a candidate for the therapy of these lymphoproliferative disorders.

The ointment preparation of hexadecylphosphocholine [15], an alkylphosphocholine, allowed us to explore this class of new drugs in a phase I/II clinical trial in patients with cutaneous lymphomas.

Patients, Materials and Methods

Patients

The study protocol was approved by the ethical committee of the University of Würzburg. Written informed consent was obtained from each patient. Patients with histologically proven cutaneous lymphoma were included if they did not require systemic therapy. There had to be measurable or evaluable involvement of the skin which was documented by photography. Patients with leukopenia, anemia, severe cardiac, hepatic or renal disease were excluded. No systemic chemotherapy was allowed during the trial and 4 weeks prior to the entry of the study. Topical steroids and photochemotherapy (PUVA) of indicator lesions were not allowed for a period of at least 2 weeks before study entry. Prior to and during the application of topical hexadecylphosphocholine standard laboratory parameters were monitored. In addition, serum was collected for determination of circulating levels of hexadecylphosphocholine. In the case of a clinical complete response, a biopsy was performed to verify the histological appearance of hexadecylphosphocholine induced tumor regression.

Hexadecylphosphocholine Ointment (Miltefosine)

The ointment used in this trial was supplied by ASTA Pharma AG, Frankfurt, FRG. It contains 6% of hexadecylphosphocholine and has been used for topical treatment of skin metastasis of breast cancer in prior studies [15].

Area of Treatment

For the first 6 patients the treated area was limited to 200 cm^2. In subsequent patients, the treated area was extended stepwise up to a maximum area of 3,200 cm^2 in the present group of patients.

Dosage and Duration of Therapy

During the first week the ointment was applied once daily, in the following 7 weeks twice daily in the defined area. The dosage was approximately one drop per 10 cm^2 lesion area. The initial duration of therapy was 8 weeks. In the case of partial or minor response continuation for additional 4 weeks was allowed.

Criteria of Response

A complete response (CR) is defined as the complete disappearance of all lesions in the treated area for at least 4 weeks. A partial response (PR) is defined as equal to or greater than a 50% decrease of all lesions in the treated area, lasting for at least 4 weeks, without any appearance of new lesions. Minor response (MR) is a regression of cutaneous lesions by at least 26% and maximal 49%. Stable disease (SD) is defined as no increase or decrease of cutaneous lesions by more than 25%. Progressive disease (PD) is at least a 26% increase of measured lesions. In case of clinically diagnosed CR, histological verification was requested.

Histological Monitoring

In each patient, a biopsy was taken to confirm the clinical diagnosis. Standard histological staining (HE, PAS, Giemsa) was used in formalin-embedded tissues. Snap-frozen biopsy specimens were stained with a panel of monoclonal antibodies (APAAP technique). An additional biopsy was taken in patients demonstrating a complete clearing of treated cutaneous lesions.

Results

Response

Fifteen patients with histologically proven cutaneous lymphomas were treated. Nine were male, 6 were female. The mean age was 60 years (range 32–86 years). Eight patients presented with mycosis fungoides of different stages (table 2a). Five patients had cutaneous B cell lymphoma (table 2b); staging revealed no extracutaneous tumor manifestation in 4 of them. One patient with low-grade B cell lymphoma (lymphocytic immunocytoma) presented nodular infiltrates of a chronic lymphocytic B cell leukemia without anemia or thrombocytopenia (fig. 2a). In addition, 2 patients with lymphomatoid papulosis were treated.

In the group of cutaneous T cell lymphoma patients, 7 patients were evaluable for response. Two complete (fig. 1a/b), two partial remission, two stable disease and one progression were observed, clinically (table 2a). In the 5 patients with B-cell lymphoma, one complete (fig. 2a, b), three partial remissions and one stable disease were seen (table 2b). Both patients with lymphomatoid papulosis (table 2c) showed a complete clearing of the lesions (complete remission).

Table 2. Patients' characteristics

Patient	Age Years	Sex	Diagnosis	Stage [3]	Response
a Cutaneous T cell lymphoma patients					
K.J.	53	M	MF	IA	SD
G.A.	76	M	MF	IVA	SD
G.C.	86	M	MF	IA	CR
T.H.	60	M	MF	IIA	PD
S.J.	32	M	MF	IA	n.e.
N.E.	50	M	MF	IA	PR
L.H.	32	M	MF	IA	CR
Z.G.	54	F	MF	IIB	PR
b B cell lymphomas					
R.P.	77	M	cb-cc		PR
K.H.	79	M	cc		CR
S.M.	77	F	cb		SD
W.M.	41	F	cb-cc		PR
S.A.	74	F	cb-cc		PR
c Lymphomatoid papulosis					
L.A.	77	F	LP		CR
W.E.	36	F	LP		CR

MF = Mycosis fungoides; n.e. = not evaluable; cb = centroblastic; cc = centrocytic; LP = lymphomatoid papulosis.

Although the calculation of an overall response rate is critical due to limited number and the heterogeneous group of patients, we indicate an objective response rate (partial and complete response) of 71% (10/14) for the cutaneous lymphoma patients treated in this study.

One patient presented a keloid at a biopsy site before starting hexadecylphosphocholine therapy. In this patient, an impressive regression of the keloid was observed during the topical treatment.

Toxicity

The topical application was tolerated without any local or systemic side effects in 4 patients. A slight erythema with fine scaling and a discrete atrophy of the skin was a common finding during the last 4 weeks of therapy in 9 patients. However, 2 patients presented a striking, sharply demarcated erythema in the treated area. They reported local itching and 'burning'. In these patients, the therapy was discontinued. These side effects were observed in the intertriginous areas (elbow and groin) in both patients suggesting an occlusive effect on the topical medication. Symptoms re-a 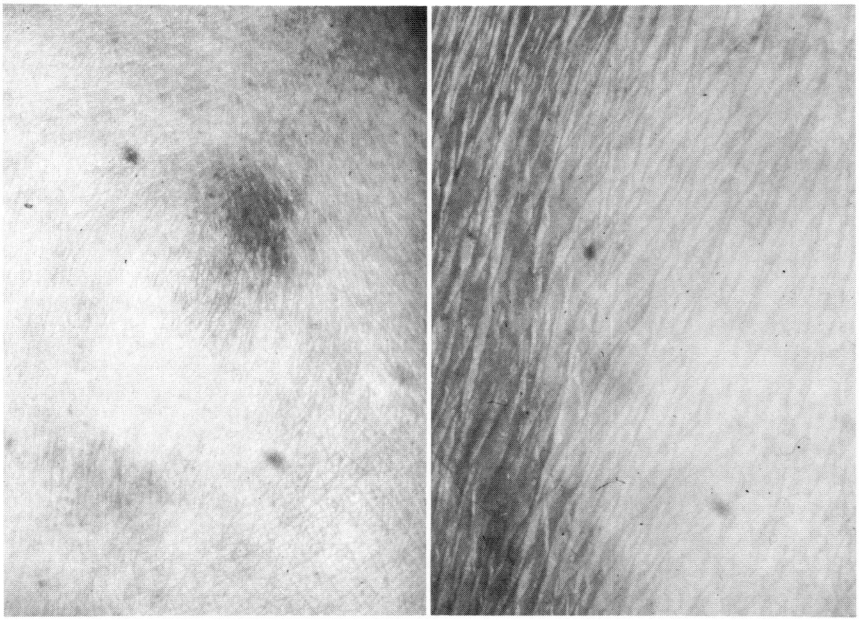 b

Fig. 1a, b. Clinical appearance of a mycosis fungoides lesion prior to and after 8 weeks of miltefosine treatment (complete response) which shows wrinkling and the thin, almost transparent aspect of the skin with postinflammatory hyperpigmentation posttreatment (*b*).

solved within a few days while the local inflammation was treated by topical steroids. It should be emphasized that the local application of hexadecylphosphocholine did not have any impact on hematopoesis, renal or hepatic function documented by standard laboratory tests. Furthermore, none of the patients noted any subjective complaints.

Histological Monitoring

In all patients treated in this study, pretreatment biopsies were taken to confirm the diagnosis. In 5 patients with complete remission and 1 patient with stable disease, a second biopsy was taken from the treated area after 8 weeks of treatment. In all biopsies taken from a treated area, a thinning of the epidermis was observed suggesting an inhibitory effect on epidermal cell proliferation.

In the patients presenting a complete response clinically, the histological section showed a significant decrease of the number of infiltrating lymphocytes in the epidermis and the upper dermis. However, the infiltrates *a* 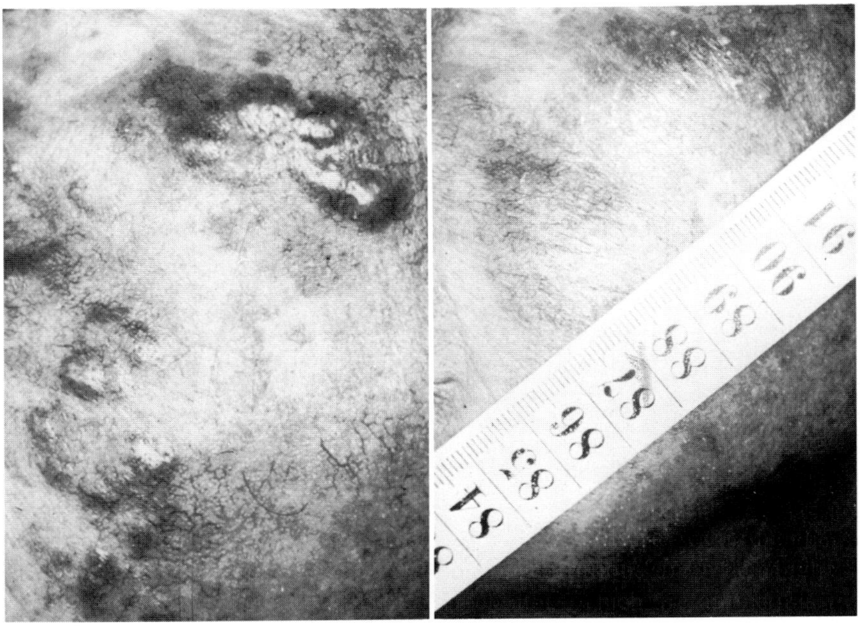 *b*

Fig. 2a, b. Cutaneous eruptions of a patient with chronic lymphocytic B cell leukemia prior to and after 8 weeks of miltefosine treatment (complete response). Note postinflammatory hyperpigmentation and continuing teleangiectasias.

in the deeper dermis were still present (fig. 3a/b). Thus, the clinical CR turned out to be PR with regard to histology.

Discussion

The clinical management of cutaneous lymphoma remains a challenging problem in dermatological oncology. The low prevalence and the highly variable natural history, with episodes of spontaneous remissions and exacerbations raised the question about the optimal management of these patients. After a controversial discussion, nonaggressive, stage-adapted, sequential application of topical therapies are preferred first-line treatment modalities. In addition, the advanced mean age of these patients suggests a moderate well-tolerable therapy. The group of patients treated in our study with a mean age of sixty years is rather representative for patients with cutaneous lymphoma [16]. However, there are only a very limited number of mild therapeutic alternatives available which include topical steroids and photochemotherapy with PUVA (psoralen plus UV-A). More aggressive is

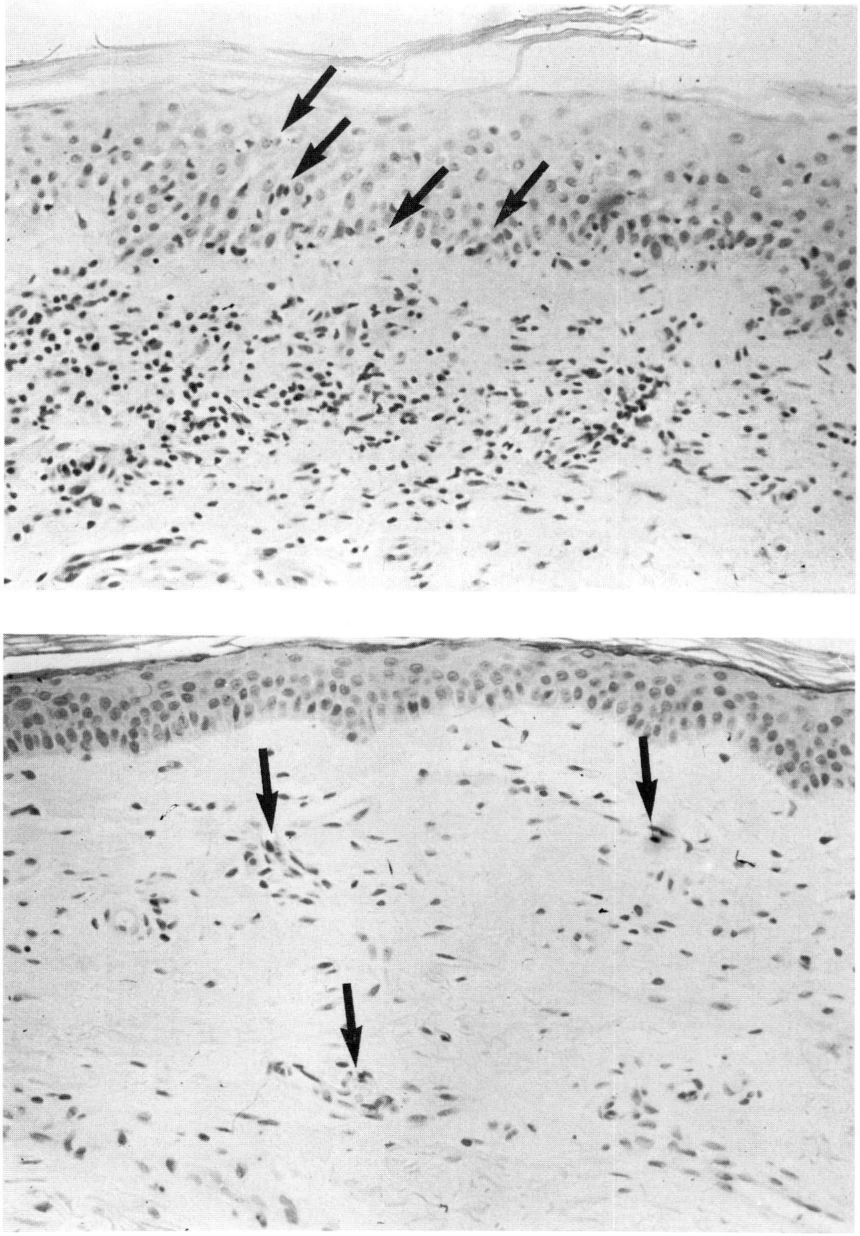
a
b

Fig. 3a, b. Histology (HE) of a mycosis fungoides lesion prior to (*a*) and after 8 weeks of miltefosine treatment (*b*). Epidermotropic mononuclear cells (*a*: arrows) have cleared. Mononuclear cells are still detectable in the deeper dermis (*b*: arrows).

the topical application of cytotoxic drugs such as carmustine, but systemic side effects such as thrombocytopenia are common [5]. New promising approaches include interferon-alpha or desoxycoformycin [10, 17]. The clinical efficacy of these drugs remains to be determined.

Therefore, it is necessary to search for new agents with low toxicity. Phospholipids are effective against a broad variety of tumor cells in vitro [13, 14, 18, 19, 20]. In the case of hexadecylphosphocholine, which was introduced as antineoplastic agent by Eibl and Unger [21], this compound shows strong and selective antitumoral activity. The topical application is feasible and provides high tolerabililty [15]. We observed no systemical side effects during the topical application which was limited, however, to an area of 20–40 cm^2 in most of the patients.

In 2 patients (13%), local inflammatory reactions caused the discontinuation of the therapy. Both events occurred early (3 and 4 weeks after initiation) during the study period in intertriginous areas. This point suggests a drug-related toxicity which was probably enhanced by an occlusive effect due to the localization of the lesions. Allergologic testing to exclude sensitization, however, was not performed.

Histological demonstration of residual infiltrates in the deeper dermis of clinically responding lesions indicates that cytotoxicity against the infiltrate seems to be restricted to the epidermis and the upper dermis. Limited skin penetration, clearance of the drug via dermal blood vessels, and local metabolic deactivation might explain this restriction of activity.

Formulations with increased skin penetration may be expected to enhance local bioavailability and to facilitate cytotoxicity against deeper infiltrates. The determination of hexadecylphosphocholine serum levels will cast light on this point, especially during treatment of larger areas.

The preliminary results reported here are promising. Miltefosine appears as an active agent against the infiltrating cells in cutaneous lymphomas. Due to the biological property of epidermotropism of these lymphoproliferative disorders, a topical application of cytotoxic drugs is suitable. This study is ongoing and it is reasonable to assume that it will provide data to understand in more detail the therapeutic efficacy of miltefosine in cutaneous lymphomas as well as in other neoplasias.

References

1 Lennert K, Feller AC: Histopathologie der Non-Hodgkin-Lymphome. Berlin, Springer, 1990, pp 162–167.
2 Burg G, Braun-Falco O: Cutaneous Lymphomas, Pseudolymphomas and Related Disorders. Berlin, Springer, 1983.
3 Burg G, Sterry W (eds): Recommendations for Staging and Therapy of Cutaneous Lymphomas. EORTC/BMFT Project Group 1987, pp 1–14.

4 Wood GS, Strickler JG, Deneau DG, Egbert B, Warnke RA: Lymphomatoid papulosis expresses immunophenotypes associated with T cell lymphoma but not inflammation. J Am Acad Dermatol 1986;15:444–458.
5 Weiss LM, Wood GS, Trela M, Warnke RA, Sklar J: Clonal T-cell population in lymphomatoid papulosis: Evidence of a lymphoproliferative origin for a clinically benign disease. N Engl J Med 1986;315:475–479.
6 Macaulay WL: Lymphomatoid papulosis update. Arch Dermatol 1989;125:1387–1389.
7 Kaye FJ, Bunn PA, Steinberg SM, Stocker JL, Ihde DC, Fischmann AB, Glatstein EJ, Schechter GP, Phelbs RM, Foss FM, Parlette HL, Anderson MJ, Sausville EA: A randomized trial comparing combination electron-beam radiation and chemotherapy with topical therapy in the initial treatment of mycosis fungoides. N Engl J Med 1989;321:1784–1790.
8 Edelson RL: Treatment of Cutaneous T Cell Lymphoma; in van Vloten WA, Willemze R, Vejlsgaard GL, Thomsen K (eds): Current Problems in Dermatology, vol 19, Basel, Karger, 1990, pp 226–237.
9 Zackheim HS: Topical mechlorethamine and carmustine for cutaneous T-cell lymphoma. Semin Dermatol 1983;2:307–318.
10 Thestrup-Pederson K: Interferon therapy in cutaneous T-cell lymphoma; in van Vloten WA, Willemze R, Vejlsgaars GL, Thomsen K (eds): Current Problems in Dermatology, vol 19. Basel, Karger, 1990, pp 258–263.
11 Kirby JD, Munro DD: Steroid induced atrophy in an animal and human model. Br J Dermatol 1976;94(suppl 12): 111.
12 Roenigk HH Jr, Caro WA: Skin cancer in the PUVA-48 cooperative study. J Am Acad Dermatol 1981;4:319.
13 Fleer EAM, Kim DJ, Eibl H, Unger C: Cytotoxic activity of lysophosphatidylcholine analogues on human lymphoma Raji cells. Onkologie 1990;13:295–300.
14 Schick HD, Berdel WE, Fromm M, Fink U, Jehn U, Ulm K, Reichert A, Eibl H, Unger C, Rastetter J: Cytotoxic effects of ether lipids and derivatives in human nonneoplastic bone marrow cells and leukemic cells in vitro. Lipids 1987;22:904–910.
15 Unger C, Eibl H, Nagel GA, von Heyden HW, Breiser A, Engel J, Stekar J, Peukert M, Hilgard P, Berger M: Hexadecylphosphocholine in the topical treatment of skin metastases: A phase I trial. Contrib Oncol 1989;37:219–223.
16 Weinstock MA, Horm JW: Population-based estimate of survival and determinants of prognosis in patients with mycosis fungoides. Cancer 62:1658–1661.
17 Dang-Vu AP, Olsen EA, Vollmer RT, Greenberg ML, Hershfield MS: Treatment of cutaneous T-cell lymphoma with 2′-desoxycoformycin (pentostatin). J Am Acad Dermatol 1988;19:692–698.
18 Fleer EAM, Unger C, Kim DJ, Eibl H: Metabolism of ether phospholipids and analogs in neoplastic cells. Lipids 1987;22:856–861.
19 Kosano H, Takatani O: Inhibition by an alkyl-lysophospholipid of the uptake of epidermal growth factor in human breast cancer cell lines in relation to epidermal growth factor internalization. Cancer Res 1989;49:2868–2870.
20 Verdonck LF, Witteveen EO, vanHeugten HG, Rozemuller E, Rijksen G: Selective killing of malignant cells from leukemic patients by alkyllysophospholipid. Cancer Res 1990;50:4020–4025.
21 Eibl H, Unger C: Hexadecylphosphocholine, a new and selective antitumor drug. Cancer Treat Rev 1990;17:233–242.

Reinhard Dummer, Department of Dermatology, University of Zürich
Medical School, Gloriastrasse 31, CH-8091 Zürich (Switzerland)

Subject Index